WEARABLE ANDROID™

Android Wear & Google Fit App Development

SANJAY M. MISHRA

LIBRARY
MILWAUKEE AREA
TECHNICAL COLLEGE
MILWAUKEE CAMPUS
700 West State Street
Milwaukee, Wisconsin 53233

WITHDRAWN

005.71
A574mi
2015

WILEY

Copyright © 2015 by John Wiley & Sons, Inc. All rights reserved

Published by John Wiley & Sons, Inc., Hoboken, New Jersey
Published simultaneously in Canada

No part of this publication may be reproduced, stored in a retrieval system, or transmitted in any form or by any means, electronic, mechanical, photocopying, recording, scanning, or otherwise, except as permitted under Section 107 or 108 of the 1976 United States Copyright Act, without either the prior written permission of the Publisher, or authorization through payment of the appropriate per-copy fee to the Copyright Clearance Center, Inc., 222 Rosewood Drive, Danvers, MA 01923, (978) 750-8400, fax (978) 750-4470, or on the web at www.copyright.com. Requests to the Publisher for permission should be addressed to the Permissions Department, John Wiley & Sons, Inc., 111 River Street, Hoboken, NJ 07030, (201) 748-6011, fax (201) 748-6008, or online at http://www.wiley.com/go/permissions.

Limit of Liability/Disclaimer of Warranty: While the publisher and author have used their best efforts in preparing this book, they make no representations or warranties with respect to the accuracy or completeness of the contents of this book and specifically disclaim any implied warranties of merchantability or fitness for a particular purpose. No warranty may be created or extended by sales representatives or written sales materials. The advice and strategies contained herein may not be suitable for your situation. You should consult with a professional where appropriate. Neither the publisher nor author shall be liable for any loss of profit or any other commercial damages, including but not limited to special, incidental, consequential, or other damages.

For general information on our other products and services or for technical support, please contact our Customer Care Department within the United States at (800) 762-2974, outside the United States at (317) 572-3993 or fax (317) 572-4002.

Wiley also publishes its books in a variety of electronic formats. Some content that appears in print may not be available in electronic formats. For more information about Wiley products, visit our web site at www.wiley.com.

Android is a trademark of Google Inc.

Oracle and Java are registered trademarks of Oracle and/or its affiliates. Other names may be trademarks of their respective owners.

Library of Congress Cataloging-in-Publication Data:

Mishra, Sanjay.
Wearable Android™ : Android wear & Google Fit app development / Sanjay M. Mishra.
pages cm
Includes bibliographical references and index.
ISBN 978-1-119-05110-7 (pbk.)
1. Wearable computers. 2. Mobile apps. I. Title.
QA76.592.M57 2015
004.167–dc23

2015011300

Cover Image courtesy of iStockphoto © ava09

Set in 10/12pt Times by SPi Global, Pondicherry, India

Printed in the United States of America

1 2015

Contents

About the Author

Sanjay Mahapatra Mishra began programming in C on various flavors of Unix in the early 1990s. By the late 1990s, he started appreciating the Linux operating system while also learning and using the Java® programming language.

Over the years, he has developed diverse software systems spanning Web applications and services, messaging, VOIP, telephony, NoSQL databases, as well as mobile and embedded platforms.

He has worked for companies such as Intertrust, Eyecon Technologies, CallSource, nVoc (formerly Sandcherry, Inc.), and Starz Entertainment Group.

Sanjay has a deep interest in and appreciation of C, Java, Linux, GNU, and open-source platforms. He possesses five Sun Microsystems Certifications since 1998 as Java programmer, Java developer, Java platform architect, Java enterprise architect, and Java Web service developer.

Sanjay earned a bachelor's degree in electrical engineering from the University of Poona (Pune) in India, and has a Google+ profile at https://plus.google.com/+SanjayMishra369/.

About This Book

Wearable computing is the paradigm that entails lightweight, miniature computers that are worn much like clothing such that the user and the computer can interact at any time, as needed. "**Wearable**" is short for wearable computing device. Almost every day, the consumer, technology, and business news tell us about new and innovative wearable technology products such as smart watches, fitness sensors, smart shirts, belts, contact lenses, and more. We live in exciting times, because "wearables" are poised to find a useful and interesting place in our daily lives. In the long run, wearable technology shows potential in diverse arenas ranging from consumer, fitness, home automation, work, and more. A few of us modern human "pioneers" have already commenced to find value in wearables. Much like the motor car long ago and the smartphone in recent memory, many innovations start out as being "unnecessary" but convenient; but before long, some catch on and even reach that tipping point after which they are perceived as "necessities." Wearable technology is an interesting intersection of fashion, fitness, efficiency, productivity, and more. A diversity of *Android Wear* and *Google Fit* devices from a diversity of major manufacturers and brand names have commenced to arrive in the mass consumer marketplace. Consumers are likely to find *Android Wear* and *Google Fit*-based devices and associated Apps engaging and exciting. Software developers will likely find developing Apps for *Android Wear* and *Google Fit*, exciting and challenging in about equal measure.

This Book

This is an "introductory" book on the "new and future looking" topic of wearables in the Android™[1] and Google ecosystem. This is a technical book on wearable Computing and application software development, specifically for the ***Android Wear*** and ***Google Fit*** platforms, which were both released in 2014.

[1] Android is a trademark of Google Inc.

Target Audience

This book has been written for a range of reading audiences including wearable enthusiasts, technologists, and software developers. The hands-on-development sections covered in this book are particularly aimed at *Android* and *Java* software developers who are interested in *Android Wear* and *Google Fit* App development. Prior experience with the Java programming language is somewhat of a prerequisite for engaging with the substantial development and hands-on sections in this book. Prior experience with Android development is ideal; yet this book does concisely cover the basics of Android software development, including the setting up and configuration of an ***Android 5 (Lollipop)*** development environment from scratch. It covers the basics of Android platform and also lists resources needed for deeper exploration in that arena. This book will help readers understand the new ***Gradle*** and ***Android Studio***-based build system.

What This Book Covers

This book covers relevant history and background about the general subject of wearable computing, before heading into the world of Android software development. Wearables represent a unique category of devices; and therefore, a distinct approach to software development and interaction design is applicable.

Many developers, including myself, can sometimes be quite impatient about diving right into installing the relevant development tools and commencing writing software from the get-go. Yet, the uniqueness, newness, and novelty of wearables, as well as the fast-paced evolution in the arena of the consumer's computing ecosystem, make the case of adequate coverage of the background and theory. In general, neither technology nor a useful consumer software application exists in a vacuum or silo. A useful consumer application typically needs to factor in and leverage the overall ecosystem for its user facing functionality as well as its system-level architecture. This book progressively covers the history, core concepts, and background on wearable computing—as a foundation for understanding the unique aspects of wearable application design and development. It covers many recent developments in the overall ecosystem of personal computing, cloud-based computing, and intelligent personal assistant-based technologies, as these have some direct or indirect relevance for designing and developing wearable applications.

This book covers the ***Android Wear*** and ***Google Fit*** platforms in the Google ecosystem and includes the setting up of a suitable development environment and getting connected to hardware devices in order to write your first applications targeted for these platforms. This book is based on and covers the latest version of the Android platforms at the time of writing, namely, Android 5 (Lollipop).

This book provides a brief coverage of the Android SDK and the new Android 5 build system, which is based on **Android Studio 1.0** IDE and *Gradle*. *Android Studio* is derived from the leading *IntelliJ IDEA*® IDE from JetBrains®. *Android Studio* is available at the Android developer website: https://developer.android.com/sdk. *Gradle* is a cross-platform project build tool in the same vein as *Ant* or *Maven*—which are tools that developers typically use for building software projects.

What This Book Does Not Cover

This book does **not** attempt to provide any comparative analysis of wearable offerings from outside of the Android and Google ecosystem; nor does it cover or acknowledge the existence of such competing offerings—this is by no means a reflection on the merits of other platforms and their offerings.

This book does **not** cover *Google Glass*™, which is a head-mounted display developed by Google and available to consumers since 2014 under an "explorer" program. *Google Glass* is a "wearable" device and platform that is distinct from the *Google Fit* and the *Android Wear* platforms.

The base Android platform SDK is rather elaborate, and this book provides a brief overview and some useful links and resources on basic Android development. While this book has two chapters dedicated to the Android platform and SDK, this book may **not** adequately serve as an independent, stand-alone reference book on the entire Android platform and SDK.

How This Book Is Structured

This book is sectioned into five parts and has ten chapters.

Part I provides an introduction to **wearable computing** including background, history, and theory. It covers diverse topics and concepts some of which potentially influence wearable application and interaction design.
Part II covers the Android platform from the ground up including its relationships with **Linux** and **Java**. It also covers the setting up of **Android 5 (Lollipop)** development environment using the new Android build system and **Android Studio 1.0**. It also covers the topics of interdevice communication and device discovery in a multidevice world.
Part III covers the **Android Wear** platform and **API**, as well as the setting up of an Android Wear device for writing *Android Wear* Apps.
Part IV covers the **Google Fit** platform and **API**, including setting up of a fitness sensor device for writing *Google Fit* Apps.
Part V provides a brief overview of some areas of applicability of wearable technology.

Hardware and Software Requirements

The *Android SDK* and *Java SDK (JDK)* are available for all the major operating system (OS) platforms. The sample code and Apps developed for this book are OS agnostic. The hands-on steps, labs, and sample code for this book (and for that matter this book in its entirety) were written and developed on **Ubuntu**—a **Linux** distribution from **Canonical Ltd**.

Ubuntu is free, fun to work with, and especially useful for Android development. Android is a Linux-based OS under the covers and therefore shares some OS concepts and equivalent commands with other *nix platforms. Therefore, getting familiar with Linux as your development platform for Android software development is aligned with attaining, in the long run, a deeper understanding of Android. The use of a *nix (Unix family) OS is suggested but is optional. Any Linux distribution or Mac® OS X makes excellent choices for Android App development. For one, you will not need to install an USB driver for each Android device model that you develop and test on—which typically happens to be the case for a MS Windows®-based Android development environment.

The hands-on labs and sample code in this book will be truly useful after acquiring at least an *Android Wear*/smart watch device and optionally also a Bluetooth low energy (LE) fitness sensor device of your choice. Real hardware devices are essential in order to get a sense of their real-world characteristics and behavior. Wearables are, after all, fundamentally about real-world interactions and behavior. Some software development may be possible using emulators and virtual devices; they can be useful to some degree, as they can provide you an indicative and approximate sense of the device's attributes, during the early stages of development.

At the time of writing, an Android Wear smart watch can be purchased for less than US $200. Many *Android Wear* devices have fitness sensors for the heart rate and step counting. You may choose an Android Wear smart watch that has more fitness sensors. You may also acquire a Bluetooth smart (LE) fitness sensor device that supports a standard Bluetooth LE GATT profile. The cost of a Bluetooth smart (LE) heart rate monitor can be less than US $60.

The Google Play Store™ (https://play.google.com/store), which sells Android-related hardware, is a perfect source for purchasing *Android Wear* devices. In case you would like to purchase a Bluetooth smart (LE) fitness peripheral device, New Egg (www.newegg.com) or Amazon www.amazon.com can be helpful.

The source code in this book has been developed using the following devices:

Samsung Gear Live	(An Android Wear device)
Zephyr HxM Smart Heart Rate Monitor	(A Bluetooth smart/LE device)

It is not necessary that you acquire devices identical to the above.

Usage of Terms

The term "***App***" or "*app*" is used often in this book and means software *Application. App* is already a dictionary word, with the meaning of *Application.* And the meaning of *Application* in this book is in the context of software Application. The terms *App* and software *Application* have been used interchangeably.

The term "wearable," also used frequently in this book, refers to a wearable computing device and/or a wearable application.

Conventions

The following are the typographical conventions used this book:

Bold—has been used when introducing a major term or to emphasize a term.
Italic—has been used to indicate literal terms such as *Home* icon or *Settings* as well as terms that go together such as *Google Fit* as one unit. *Italic* has also been used for class and package names and code snippets.
Italic bold—has been used to emphasize terms as covered in the Italic section, especially at the time of their initial introduction.
`Constant width`—has been used for commands that are to be typed literally.

Diagrams Used in This Book

Some of the diagrams used in this book are covered by the **Creative Commons License** and have been attributed accordingly to their original creators. Still, other diagrams used in this book are from the **public domain**. The rest of the diagrams used in this book have been created by the author. The technical software class diagrams included in this book were created using MagicDraw© from www.nomagic.com. These class diagrams are somewhat informal and do not follow strict UML notation—as they include additional method details, comments, and such.

Third-Party, Online References

The third-party, online references and links listed in this book may change over time and are not in the control of the author or the publisher. Despite this shortcoming, they have been listed due to their relevance to the topics covered in this book.

Website

This book has one dedicated website, with an index to all the online resources associated with this book. This website has two domain names: wearableandroidbook.com and wearbook.io for convenient access. There is trend toward using *.io* in the domain name to represent input/output (I/O).

Source Code

The entire source code associated with this book is available online at the aforementioned website. The source code has not been included in the contents of this book, other than as nominal code snippets.

Errata

I have made every effort to proofread and verify every aspect of this book as much as possible. Several book reviewers have graciously read and validated various aspects of this book's contents. Nonetheless, should any errors, typos, or ambiguity be detected after publication, the errata will be available at the aforementioned website.

Trademarks and Copyrights

Android, *Google Play Store*, *Dalvik Virtual machine*, *Google Glass*, *Nexus*, *Open Handset Alliance*, *ChromeOS*, and *ChromeBook* are the registered trademarks of Google, Inc. *Android Wear* is a version of Google's Android operating system designed for smartwatches and other wearables. *Google Fit* is a health-tracking platform developed by Google for the Android operating system. *Ubuntu* is the registered trademark of *Canonical Ltd.* All other trademarks are the property of their respective owners.

References and Further Reading

Creative Commons License. http://creativecommons.org/licenses/
Public Domain. http://en.wikipedia.org/wiki/Public_domain

Acknowledgments

I am grateful for the casual and noncompetitive home and school environment during the early years of my life, thanks to my parents Sabita and Prafulla and also my teachers at Loyola High School in the city of Poona (Pune), India.

I am also grateful for the influence of and encouragement from several supervisors, coworker, and friends over the years: Suresh Joshi at the Software Engineering and Design Company, India, and Manny Bhangui at Citibank, India, had great technical insight and style, which inspired me during the early years of my professional work experience in the 1990s. Maureen A. McGee and Bobbie Pitzner both at AT&T in Short Hills, NJ, were highly encouraging when I was a newly arrived immigrant in the United States. Jeff Lutz at Boldtech Systems in Colorado was highly supportive professionally and personally during the brief period of our working together and beyond. Jon Ford at Sandcherry/nVoc and Linda Gonzalez at the Starz Entertainment Group were both in their own ways, highly insightful and encouraging, during my time at these respective companies. I would also like to thank Richard Steel, Uday Natra, Martin Wills, Li Wang, Chris Butler and Rob Nevitt for their collaboration and friendship.

I am especially grateful to Nathan Blair, for his meticulous and valuable feedback in reviewing this book. Nathan wrote his first program in BASIC on a TRS-80 under the guidance of his grandmother and quickly discovered that programming and computers were his passion. Since then, he has worked on a wide variety of platforms and languages. Nathan has bachelor's and master's degrees in computer science and currently lives in the Denver area in Colorado.

I would like to thank Rudi Cilibrasi for the inspiration and insight that I gained from our long conversations during the year 2012. I would also like to thank Rudi very much for his valuable suggestions and the public recommendation for this book. Rudi Cilibrasi is a computer scientist who explores math, machine learning, and networking through programming and amiable interactions with friends. Rudi develops phylogenetic tree reconstruction algorithms based on mitochondrial DNA and machine translation algorithms for natural human languages based on the World Wide Web.

I would also like to thank Franz Zemen for his valuable feedback on the initial section of this book.

Last but not the least, I would like to thank my editor Brett Kurzman and the rest of Wiley team for their valuable trust and effort in making my first book possible.

Part I Wearable Computing: Introduction and Background

This section provides history and background on Wearable computing. It includes a range of topics including the human–computer interaction paradigm, the spatial scope of computing, ubiquitous computing, and so on. Wearable devices represent a unique device form factor, and Wearable applications somewhat require a distinct interaction paradigm. Developing applications for Wearable entails some fundamental differences in the interaction and design compared to other platforms such as phone, Web, and desktop. Wearables typically coexist in an ecosystem of cloud-based computing and a multitude of devices that a given user may interact with.

Wearable Android™: Android Wear & Google Fit App Development, First Edition. Sanjay M. Mishra.
© 2015 John Wiley & Sons, Inc. Published 2015 by John Wiley & Sons, Inc.

Chapter 1 Wearables: Introduction

1.1 Wearable Computing

In general, a computer or computing device is characterized by the presence of a central processing unit (CPU) within it. The CPU is the crucial hardware that carries out the instructions of computer programs. **Wearable Computing** is the paradigm that entails lightweight, miniature computers that can be worn on the body such that the user and the computer can interact at any time as needed, with minimal overhead and impact on the user's real-world physical activities. Examples of such real-world physical activities are gardening, jogging, rafting, carrying a child, walking a dog, and so on. It can be harder and inconvenient to engage in many such real-world physical activities while also holding a phone or having to bring it out of the pocket or handbag frequently.

1.2 Wearable Computers and Technology

A **Wearable Computer** is a body-borne, miniature computing device, which the user has opportunity for constant access to and interaction with—with minimal impact to the user's real-world activities. Wearable Computers have historically been used for the last few decades in niche and specialized segments such as space, military, academic, medical, industrial, and so on. Wearables have also been the subject of academic research since decades. Many of the technological innovations from the academic and niche arenas are starting to be seen today in the nascent consumer Wearable segment.

Wearable Android™: Android Wear & Google Fit App Development, First Edition. Sanjay M. Mishra.
© 2015 John Wiley & Sons, Inc. Published 2015 by John Wiley & Sons, Inc.

1.3 "Wearables"

Wearable Computers or simply "***wearables***" are today no longer limited to the abovementioned niche segments; they have commenced to make their way into the mass consumer market. Wearables are available to consumers in various shapes and forms including smart watches, clothing, belts, shoes, jewelry, athletic and fitness sensors, and so on.

The calculator watch and similar products introduced in the 1980s may be considered to be instances of simple wearables. Wearables in today's world can be quite sophisticated due to the synergistic integration of various information such as the user's current contextual information or context with the Internet cloud-based intelligent agents. Mobile devices and wearables to a greater extent can provide valuable signals from which the user's "context" can be inferred. This real-world "context" refers to where the user is currently located, what the user is currently engaged in, and so on.

Wearable technology and modern human–computer interaction trends aim to make computing less intrusive on the user's real-world experience. Today, value can be derived, not from computing devices in isolation but rather from the synergistic combination and collaboration between devices and sensors in a networked and "ubiquitous" computing ecosystem. Ubiquitous computing is the concept wherein computing is accessible everywhere and at all times, via any device.

Much like Bluetooth headsets reduced the intrusiveness of smartphones while having phone conversations and simultaneously engaging in various activities, wearables such as smart watches aim to make it easier for users to engage in their diverse real-world activities while simultaneously maintaining "light," "glanceable" interactions with the online digital world.

Wearables have commenced to make an entry into the mass consumer market, due to the convergence of numerous factors. The modern human has now commenced to wear one or more computing devices on their person that are always on and ready and close at hand. This is a trend that is unlikely to go the way of some outdated fashion, anytime soon. The implications are huge, and the applications of Wearable technology have tremendous potential. Much like the motor car of long ago, and the smartphone in more recent memory, many innovations start out as being "unnecessary" but convenient; but before long, some catch on and even reach that tipping point after which they are perceived as a "necessity."

Wearable technology lies at this interesting intersection of fashion, fitness, home automation, efficiency, productivity, and more. Some of the limitations of the smartphone in terms of their intrusiveness toward the user's real-world activities make the case for wearable devices such as smart watches.

1.4 The word: "Wearables"

The word "wearables"—short for Wearable computer and technology—has been used for several decades mostly in academic and technical publications. Today, the arrival of devices to the consumer market has started to make it a commonly used word. Currently, the dictionary word "wearable" is an adjective meaning "capable of being worn." In this book, we will use the term "wearable" as a noun to denote a wearable computing device. The chances are that once this word gathers adequate mass usage—sooner or later and likely sooner than later—the major dictionaries of the English language will begin to acknowledge the use of this word "wearable" as a noun to denote the concept of "a wearable computer or device." "Wearables" are thus computing devices that are intended to be convenient to wear and comfortable to interact with, while we go about our choice of real-world activities.

1.5 Wearables and Smartphones

Wearables are typically not a replacement for smartphones or tablets—rather wearables typically complement and augment smartphones and tablets. Wearables are somewhat of a natural progression and extension of the smartphones and the useful smart "Apps" that reside and run on them, which have become an indispensable part of our daily lives. Some smartphone Apps have adequately demonstrated their usefulness and ability to serve as our own intelligent personal agent, always ready and available to help us in the many dimensions of our daily toil and strife of work and family, fitness and health, entertainment, education, and more.

Wearables as do smartphones often have sensors, which can help in determination of the user's current context. However, wearables—by virtue of being worn on the person—have more intimate sensor access including biological parameters such as heart rate, skin conductivity, body temperature, and so on, thereby making them useful for fitness and productivity applications and so on.

1.6 Wearable Light, Glanceable Interactions

Wearables support the ideal that users can more easily continue to pay adequate attention to their physical activities and environment, while also keeping up to date with the online world via lightweight, minimally intrusive interactions. Wearables are intended to help us engage better with our real-world activities that tend to change from moment to moment, in free and full flow. Wearables aim to make it easy for you to keep in touch with the physical world and environment and also be on top of those important, informational electronic updates, acknowledgments, and lightweight actions that need to be performed in real time.

1.7 Smartphone Dependency, Inconveniences

Fundamentally, we as consumers use personal computing devices because we derive some value from them. At the same time, using any computing device tends to distract and detract from our real-world activities. The more we recognize and appreciate the benefits of our smartphones and Apps that run on them, the more they become an integral part of our daily lives; and the more we tend to experience the inconvenience, overhead, and inelegance of having to frequently dig our phones out of our pockets and handbags or holding our phones in our hands for extended periods of time and under inconvenient circumstances—such that our almost perceptual use of our phones can tend to interfere with our various real-world activities.

The greater our need to keep connected with the networked world, for reasons of family, work, entertainment, and more, the more we are likely to benefit from a more elegant and less intrusive "wearable" model of the human–computer interaction. The wearable model aims to reduce the distractive and constraining effect on the user in "the here and now."

1.8 Wearable Interaction

The more trivial the nature of an electronic interaction, the more likely that the wearable will suffice. The more complex your electronic activity or task (say, something substantial such as writing a marketing plan, preparing a report, watching a movie, etc.), the more likely that you will benefit from a larger computing device such as a smartphone, a tablet, a *Chromebook*™, or a netbook computer. Smart watches typically support simple "outbound" communication using voice and simple touch menus and simple "inbound" context-based suggestions and cards.

1.9 User's Real-world Context

The user context is a broad term that includes location awareness and real-world activity recognition. It is about where a user is and what activity a user is engaged in, at any given time. Smartphones often come with various sensors such as accelerometers, gyroscope, and so on, which Apps can access and leverage in order to make an intelligent determination of the user's real-world context such as driving, running, hiking, at work, at home, and so on. Apps have recently been trending and evolving toward a more user context aware, proactive, predictive, participatory paradigm of interaction with the user. The user's real-time context awareness is one of the foundations for intelligent agent-based applications. Wearables are uniquely qualified to provide accurate and useful insight into the users' real-world context due to their various sensors and direct contact with the human body.

1.10 Variety of Wearable Devices

A wide variety of wearables such as smart watches, fitness sensor cuffs and straps, smart contact lenses, athletic goggles, eye glasses and displays, smart headphones, helmets, smart clothing, shoes and belts, smart jewelry, and so on have become available in the consumer market. A few of the most common wearable categories are listed below.

1.10.1 Smart Watches

Smart watches are one of the predominant wearable devices in the consumer market today. Initially, the rise of the smartphone tended to make the wrist watch practically redundant. But today, the success and proliferation of the smartphone and our deepening dependence on them paves the way for smart watches, which offer a less intrusive interface. Smart watches mostly serve in the role of an extension of the smartphone. Smart watches typically have one or more mechanisms for interconnectivity such as Bluetooth LE, Wi-Fi,

USB, and so on. Bluetooth LE is the predominant mechanism for connectivity. Once paired with a phone, the smart watch can access the network.

1.10.2 Fitness Sensors

Fitness sensors are available in various configurations—some are stand-alone sensors mounted on chest straps, wrist bands, and so on. Others are integrated or embedded into other body-worn items such as watches, headphones, belts, shoes, goggles, and so on. Fitness sensors typically provide connectivity via technologies such as Bluetooth LE, Bluetooth (classic), Wi-Fi, etc.

1.10.3 Smart Jewelry

Smart jewelry is less of a separate category of wearables and more of a special case of smart watches and fitness sensors that are encased in elegant and/or expensive metal. There are a variety of smart jewelry such as bracelets, rings, necklaces, and so on that perform the function of jewelry in conjunction with the computing functions as in smart watches, fitness sensors, and activity trackers.

1.11 Android Wear and Google Fit

Android Wear and *Google Fit* are distinct and collaborative efforts by Google and numerous partners to bring smart watches and fitness sensors into the mass consumer market in a user-centric ecosystem. *Android Wear* and *Google Fit* aim to make it easier for App developers to write Apps that are portable across devices from diverse manufacturers.

Android Wear and *Google Fit* are separate but closely related platforms. A typical *Android Wear* device is the smart watch—which augments the smartphone and provides a simpler and lighter user interface that allows the user to receive notifications and address trivial online interactions, in a less intrusive manner. *Android Wear* devices typically have a simple screen and can accept voice and touch inputs. The *Android Wear* watch is conceptually an extension of the smartphone. Most *Android Wear*/smart watch devices have fitness sensors such as heart rate, step counters, and so on.

Google Fit currently works with Bluetooth LE devices such as heart rate monitor or step counter worn on the body that provides sensor data that their smartphone can access.

While the general subject of Wearable Computing certainly includes medical devices, particularly *Google Fit* is a fitness platform and explicitly excludes medical devices and medical applications. Medical devices and applications are typically regulated by the country-specific governmental agencies.

1.11.1 Device / Hardware Purchases

The subject of procurement of an *Android Wear* device has been covered in Section 6.7.2. Similarly, the subject of procuring devices for *Google Fit* development has been covered in Section 8.8. You may refer to these mentioned sections in advance, in case you would like to order suitable devices now or at any point.

References and Further Reading

http://en.wikipedia.org/wiki/Wearable_computer

http://en.wikipedia.org/wiki/Wearable_technology

http://www.forbes.com/sites/gilpress/2014/08/22/internet-of-things-by-the-numbers-market-estimates-and-forecasts

http://spectrum.ieee.org/consumer-electronics/portable-devices/wearable-computers-will-transform-language

http://en.wikipedia.org/wiki/Head-up_display

http://www.media.mit.edu/wearables/

http://www.android.com/wear/

https://developers.google.com/fit/

http://en.wikipedia.org/wiki/Accelerometer

http://en.wikipedia.org/wiki/Gyroscope

http://en.wikipedia.org/wiki/Context_awareness

Chapter 2 Wearable Computing Background and Theory

2.1 Wearable Computing History

Depending on how we define the concept of *Wearable Computing*, how long it has been around can range from hundreds of years to several decades. At some level, even a wrist watch can be considered to be a computing device—it computes the time for you and is always on, ready, and available. But the term computing device or computer is characterized today by the existence of a processor (CPU) within—which is missing in the case of, say, a mechanical watch. At the same time, the earliest computers were mechanical and electromechanical in nature. All this can make it harder to arrive at one definite answer that is widely acceptable.

Since the 1980s, the consumer marketplace has certainly seen sophisticated digital watches with scientific calculation, games, audio, and video capabilities. *Nelsonic Industries*, a US company, produced a "game" watch in the 1980s, which served both as a timepiece as well as an electronic game device. *Casio*, the Japanese electronic company, offered a wide variety of digital watch models including the *Casio CFX-400* scientific calculator watch as well as the *Casio Databank CD 40*, both of which were introduced in the early to mid-1980s (Figure 2-1).

Sophisticated, modern *Wearable Computing* has certainly been in use in military, industrial, and research and educational labs since the last few decades. For instance, headgear with displays has been used in the arena of military and space applications. Similarly, head-mounted displays have been in use by surgeons for performing advanced surgeries.

Wearable Android™: Android Wear & Google Fit App Development, First Edition. Sanjay M. Mishra.
© 2015 John Wiley & Sons, Inc. Published 2015 by John Wiley & Sons, Inc.

Figure 2-1 Casio CFX-400 watch manufactured circa 1985. Attribution: By Septagram at en. wikipedia [Public domain], from Wikimedia Commons.

Today, the overall cloud-based ecosystem, economic factors, and also the human/user expectation from mobile devices have evolved to a point where wearables have great potential in the mass consumer market.

2.1.1 Wearable Computing Pioneers

There are several individuals who may be credited as pioneers of modern *Wearable Computing*. The following is a partial listing of such computer scientists, along with brief highlights of their contributions.

Edward O. Thorpe and Claude Shannon, both MIT Ph.D. mathematicians, have been widely credited for designing the first *Wearable Computer* in the early 1960s. Thorpe pioneered the application of probability theory in various arenas including hedge fund techniques in financial markets as well as the mathematics of gambling. Thorne and Shannon developed the first *Wearable Computer* and used it as a gambling aid when playing the game of blackjack, purportedly as a purely academic exercise (at a casino in Las Vegas, Nevada, where gambling was and continues to be legal). Subsequently, laws in the state of Nevada were revised to outlaw the use of *wearables* and other computing devices to predict the outcomes in betting and gambling.

Steve Mann is a Ph.D. researcher and inventor whose work since the 1980s has contributed immensely toward modern *Wearable Computing* technologies. Mann designed a backpack-mounted computer to control photographic equipment in the early 1980s while still in high school. Mann was one of the founders of the *Wearable Computing Lab* at MIT, and he continues to be an active contributor in the field of *Wearable Computing*.

Thad Stamer, a Ph.D. researcher and professor, has been actively contributing to the field of *Wearable Computing* and the associated topics of contextual awareness, pattern recognition, human–computer interaction (HCI), and artificial intelligence since the 1990s. Stamer was one of the founders of the MIT's *Wearable Computing Project* and has a key role with the *Google Glass* project.

Edgar Matias and Mike Rucci from the University of Toronto have been credited with building a wrist computer in the 1990s. Mik Lamming and Mike Flynn at Xerox PARC demonstrated in the 1990s a wearable device, the Forget-me-not, that could record interactions with people and store them for later reference. Alex "Sandy" Pentland, a Ph.D. computer scientist, is an active faculty member at MIT Media Labs and one of the pioneers of wearable and data sciences. His work has focused on wearables, human, and social dynamics using data analytics.

As with any arena of complex technology and research, the knowledge base and ecosystem have been built based on the contributions from a multitude of dedicated researchers, professionals, enthusiasts, and hobbyists.

2.1.2 Academic Research at Various Universities

Wearable Computing has been a theoretical subject of academic research for several decades. A wide variety of pioneering work has been done at the *MIT Media Lab*, which focuses on the convergence of technology, multimedia, science, art, and design. HCI and wearable technologies are some of its core arenas of research. *Columbia University* too has pioneered work on augmented reality since the 1990s. Numerous universities continue to conduct research in the field of Wearable Computing and related fields of virtual reality and augmented reality. Much of the innovations seen in consumer market wearables today have originated from the academic research of years and decades ago.

At a fundamental level, Wearable Computing and research has a correlation with both virtual reality and augmented reality. "Virtual reality" provides the user with a simulated, unreal, or virtual sensory input.

Augmented reality on the other hand enhances the real world with additional computer-generated information such as audio, video, and other data, such that the user's current perception of reality is enhanced. Augmented reality provides additional input that overlays and co-exists with the user's real-time experience of their real-world environment. Augmented reality often uses computer vision and object recognition. Using virtual reality, NASA astronauts can "experience" being at a beach during the long periods of time in space, and get some relief from the psychological and physical demands of being in space, away from their home planet. The same astronauts benefit from augmented reality-based systems that are typically used while carrying out critical physical missions and maneuvers in space.

2.2 Internet of Things (IoT) and Wearables

The Internet of Things (IoT) may be thought of as an interconnected ecosystem wherein diverse computing devices, large and small, interconnect, collaborate, and cooperate with one another. The IoT represents the proliferation of small embedded computing devices that interconnect with the Internet. "Things" in the IoT include diverse devices such as health monitoring sensors, automobile sensors, pet biochips, indoor plant moisture sensors, home appliances, industrial and smart factories, smart transportation, agriculture-related sensors, and so on. There are estimated to be millions of IoT devices today, and according to market research and forecasts by the Harvard Business Review, 28 billion such IoT-type devices are expected to join the IoT by year 2020 (Figure 2-2).

The world of IoT spans the arenas of consumer, industrial, transportation, safety, farming, and more. Environmental monitoring, infrastructure management, manufacturing and asset tracking, inventory control, safety monitoring, medical and health care, transport systems, fleet management, resource optimization and energy management, etc. are some examples of the application of IoT technology.

A discussion on IoT often brings up the topic of wearables and vice versa. IoT devices include wearables; some of the devices, that is, "things" in the world of IoT, happen to be wearable devices. Some wearable devices may interact with non-wearable IoT devices in order to share information or instructions depending on the needs of the applications.

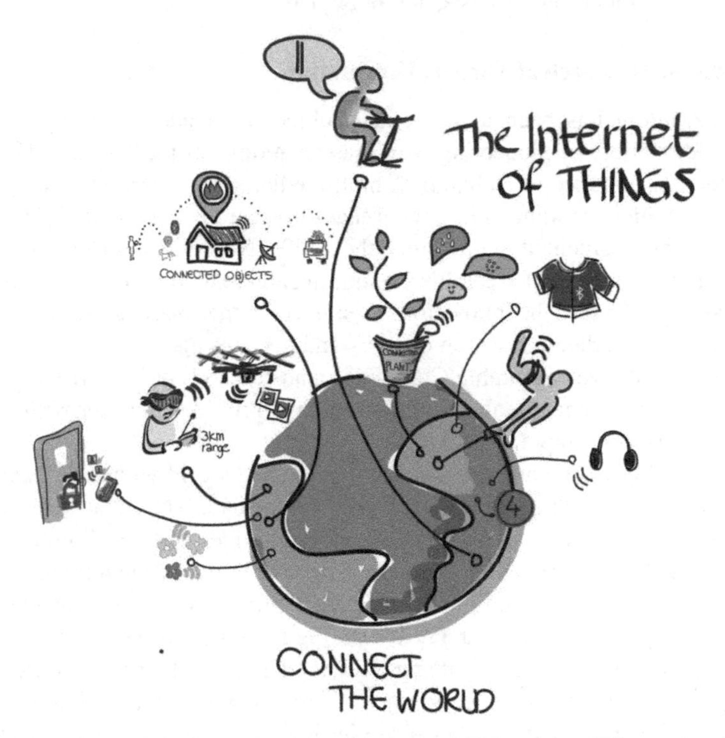

Figure 2-2 Internet of things. Attribution: "Internet of Things" by Wilgengebroed on Flicker, licensed under Creative Commons 2.0.

Many definitions of IoT devices emphasize the aspect of their direct network addressability on the Internet, as in an independent IP address directly on the Internet. In practice, however, it is not necessary for each IoT device to have independent connectivity and addressability on the Internet. This is because IoT devices typically have access to other local networks, which in turn have the capability to connect to the Internet. With the cloud serving as an intermediary, it is possible for IOT devices to connect as client devices to the Internet-based cloud endpoints and maintain long-lived connections over which they can receive data and commands (without needing to possess their independent IP address directly on the Internet). Devices worn on the user's body, for example, have access to the user's phone via Bluetooth, which in turn typically has carrier service. Smart thermostats, washers, dryers, refrigerators, sprinkler systems, etc. in a user's home connect to the local home network over Wi-Fi or wire. Devices in an industrial environment often have a local network. In all these instances, proximate devices and networks can pair, tether, and interconnect in order for the IoT-type devices to bridge the network path to the Internet and connect to Internet cloud-based endpoints as connected client devices and thereby attain two-way communication capabilities.

Internet and phone service providers typically have a business model that benefits from provisioning Internet access for individual devices including IoT devices independently—with the associated costs to the consumer. There are numerous disadvantages of having smaller devices directly on the Internet with their own IP address. For one, they are more susceptible to Internet cyberattacks and hacking. Also, it is typically more expensive to have to provision and provide direct connectivity for every IoT device independently. Traditionally, we have had devices within the local network that do not have their own Internet address, which can remain connected to the Internet all the time—as client devices. For example, personal computing devices within a local home network connect as client devices and have the ability to send and receive information.

The use of the cloud as an intermediary helps devices provide remote networked services without exposing the devices as Internet endpoints. As an example, Google Cloud printers allows you to print from anywhere over the network, yet the Google Cloud printer is connected on your local network via which it connects as a client device to Google Cloud on the Internet; it is not exposed as a device on the Internet with its own public Internet IP address, and yet it services requests to print documents over the Internet via the cloud as the intermediary.

2.2.1 Machine to Machine (M2M)

The term *Machine to Machine* is associated with and commonly used along with the term IoT. M2M and IoT are associated and have significant overlap of concepts and applications. M2M refers to the direct two-way connectivity between a computing device and a "fellow" computing device. In the world of M2M, devices typically have a network address directly on the Internet, and devices are peers on the network. M2M was originally used in specialized industrial segments including automation, instrumentation, and process control. M2M devices are generally characterized by small devices having their own IP address directly on the Internet.

2.3 Wearables' Mass Market Enablers

Today, a coincidence of many factors has paved the way for the arrival of wearables to the mass consumer market. The manufacturing economy of scale has made small powerful devices more affordable. The human dependence and appreciation of mobile computing has expanded the possibilities for wearable devices to provide additional value to the users.

2.3.1 "ARM-ed" revolution

The **ARM** family of processors power the vast majority of mobile devices such as mobile phones, tablets, and small embedded devices including the majority of the IoT devices. ARM processors also power set-top boxes, televisions, and netbook computers. The ARM family processors have typically required significantly less power compared to the *x86* family of processors. ARM processors are light, portable, and small in size. The success of mobile devices has hinged on the availability of low-cost, lightweight, compact, and power-efficient ARM processors. ARM has played a key role in the last many years in the success of affordable mobile devices that can run on battery power for hours on end.

Over 50 billion ARM processors have been manufactured as of 2014, and this appears to be only the beginning. The availability of wearable devices hinges for the most part on the ARM-based processors and system on chip (SoCs) covered in the next section. Incidentally, ARM processors have recently made an entry into the market segment of servers that reside in data centers on the cloud.

ARM Holdings plc is the British company that bears the ARM name and develops the architecture and design of processors. *ARM Holdings* licenses the technology to other companies for purposes of production and manufacturing. ARM licensees include companies such as Qualcomm, Broadcom, Marvel, Freescale, Amtel, Nvidia, Texas Instruments, NXP, ST Microelectronics, Applied Micro, AMD, Samsung, and Apple. These are the companies that make ARM processors or chips; some of these companies also manufacture and market mobile devices as well.

2.3.1.1 ARM alternatives MIPS from *MIPS Technologies, Inc.* is an instruction set that competes with the ARM family of processors. MIPS has been used by embedded and real-time Linux-based operating systems since decades. Android was ported to MIPS in the year 2009. *MIPS Technologies, Inc.* was acquired by *Imagination Technologies*, a UK-based processor R&D and licensing company that is widely known for their graphics processors. Several prominent companies such as Broadcom offer both MIPS- and ARM-based processors/SoCs for various market segments.

Intel, which is well known for its highly popular *x86* family of processors, has been working on creating low-cost and low power consuming *x86* family processors known under monikers such as "Atom," "Quark," "Edison," etc., which are aimed at the IoT and wearable market segments.

2.3.2 System on Chip (SoC)

Although the CPU is the heart of any computing device, the CPU works not in isolation but in conjunction with several other key components such as memory, graphics processing unit (GPU), audio chips, wireless radios (Wi-Fi, 4G), USB controllers, and so on. A *system on chip* (SoC) integrates the CPU with the other key components such as memory, GPU, USB controller, wireless radio, etc. into a single integrated chip.

While a computer cannot be built based on a single CPU chip alone, it can be built based on a single SoC chip. In recent years, the trend has been in the direction of SoC-based consumer devices, coincident with the proliferation of mobile devices and the success of the ARM-based processors.

The SoC has the advantage of its highly compact size, lower cost, and less wiring. SoCs have now begun to appear inside larger systems such as netbooks and even server systems. The limitation of the SoC can be that it is not conducive toward the upgrade or replacement of individual constituent components, as they have all been integrated upfront.

The mass production and availability of low-cost wearable devices hinges on the availability of SoCs in an innovative and competitive market. Today, there are well over 50 SoC manufacturers who manufacture SoCs based on ARM-, MIPS-, or Intel-based processors.

2.3.3 Human Dependence on Computing

Consumer interaction and dependence on computing started with the personal computer (PC), but it was the success of the smartphone that deepened this dependency on and appreciation of computing devices on a more engaging, intimate and somewhat "constant" basis. Now that the smartphone has become an integral part of most users' lives and the value from computing in their daily lives adequately appreciated, there is greater interest and potential value from the enhanced functions and specialized use cases that wearables can provide.

2.3.4 Smartphone extensions

In general, smartphones have tended to provide a software solution and a software-based replacement for many hardware gadgets. At first, and since well over a decade ago, the mass adoption of feature phones and later smartphones tended to cause a progressive decline in the use of hardware devices such as digital watches, alarm clocks, cameras, scanners, simple level instruments, and so much more. In a way and to some extent, the introduction of wearables into the consumer computing ecosystem today seems to run contrary to this trend. Today, our appreciation of and dependence on the smartphone paves the way for devices such as a wearable smart watch. Users experience some of the limitations of the much utilized smartphone—in terms of their intrusiveness toward the user's real-world activities—and this helps justify wearable devices, some of which aim to elegantly extend the smartphone and act somewhat like an accessory and extension of the smartphone. Still other wearables aim to provide specialized functionality such as fitness sensors and health monitoring.

2.3.5 Sensors

Many smartphones have sensors for motion, position, environment (temperature, pressure, humidity, and light), and so on. Fitness devices have sensors for fitness-related parameters such as heart rate, step counters, etc. Sensors play a key role in the world of wearables and IoT. The inference of the user's context and access to fitness readings—which tend to make wearable applications more useful to consumers—are possible due to the availability of low-cost, power-efficient sensors.

2.3.5.1 Micro-Electro-Mechanical Systems (MEMS) Sensors

Much like compact, power-efficient, and economically priced embedded computing devices are made possible in the consumer market by the availability of a wide variety of SoCs, the compact, power-efficient, and economically priced sensors are made possible predominantly by micro-electro-mechanical systems (MEMS) technology. MEMS devices have a size in the range of 20 micrometers to 1 millimeter and are made up of components that have a size in the range of 1–100 micrometers (0.001–0.1 millimeters). MEMS devices typically consist of a central unit as well as micro-sensors and micro-actuators that interact with the surroundings. MEMS devices can serve the functions of sensors and actuators. MEMS at the smaller scale merges into nano-electro-mechanical systems (NEMS). The nanoscale refers to structures in the order of 1–100 nanometers. One nanometer is one billionth of a meter.

MEMS technology is found to reside within accelerometers, gyroscopes, touch sensors, temperature sensors, humidity sensors, microphones, health and medical sensors, inkjet printers, game controllers, automobiles (dynamic stability control and tire pressure sensors), hard disks (to "park" the head when free fall is detected, in order to protect the disk and prevent damage and data loss), and much more.

2.4 Human–Computer Interface and Human–Computer Relationship

The Wearable Interaction represents a different flavor of HCI/human–computer interface. The human–computer interface as well as the human–computer "relationship" has been evolving over time.

2.4.1 Human–Computer Interface: over the years

The human–computer interface refers to the interaction between human and computer. The computer receives input data from the human via various mechanisms such as keyboard, soft keyboard, mouse, touch, gestures, speech, audio, video, vision, and so on. The computer responds with output data such as screen displays, printouts, audio, video, and more.

Until the 1970s, most of the input and output—the interaction between human and computer—was mostly based on "punch cards." A punch card is a thick paper card that encodes programs and data. Computer scientists, programmers, and operators used a key-punch, a typewriter-like device to write data that was fed to the mainframe computer via a punch card reader. This was the era of the "mainframe" computer that was used in the corporate, industrial, military, and academic worlds. This punch card-based interaction kept the use of computers limited to computer scientists, programmers, and operators and as far as possible from the consumer. At that time, consumers typically had no interaction with or direct use of such computers or any computers at all. The advent of the PC era, which began in the late 1970s and 1980s, changed that model, slowly but profoundly. Incidentally as of 2012, some voting machines in the United States reportedly used punch card based mainframe computers.

In the PC era of the 1980s through to the recent decade, the computer and the consumer interacted via the keyboard, monitors, and mouse and to some degree via speakers and microphones. In this PC era, the users' computer was a located in their home or office, perched on their desk. Soon, the laptop arrived and was easier to carry around on business and leisure.

In the post-PC world of today—the era of mobile and cloud—consumers interact with their mobile computing devices via intuitive mechanisms such as soft keyboards, touch, gestures, voice, audio, video, and so on. Computing resources on the cloud are also typically an important part of this interaction; however, these cloud computing resources are somewhat abstracted out in terms of their location, specification, and power needs. The consumer cares about the quality of experience and service but typically neither knows nor particularly cares where the cloud-based computers that serve their needs physically reside nor what their specifications of CPU, memory, etc. look like. In retrospect, the PC was not all that personal since it was not that close at hand, compared to today's mobile devices which are close at hand all day and thus are more personal.

2.4.2 Human Computer Interaction (HCI): Demand and Suggest

Two important paradigms in modern human computer interaction (HCI) design are the "Demand" and "Suggest," which are covered in this section.

2.4.2.1 Demand Paradigm In general, the HCI certainly started out in a "Demand" paradigm wherein the human demands some information or action from the computer, while the computer provides the information or performs the action in response to the "Demand."

The *Demand Paradigm* has historically been and still is prevalent in the world today. In the *Demand Paradigm*, the human is in the driver's seat and asks the computer for information, while the computer "dumbly" responds. The *Demand* model in a way underlines this somewhat of a master–slave relationship between human and computer.

Certainly, the human is the master here, and humans created computers to serve and assist them. Yet, the limitation of the *Demand* model lies in the fact that the mundane actions need to be initiated by the human every time and it is somewhat of a "manual" process for a human to have to remember to initiate something—that might be predictable. It is often predictable that a certain *Demand* is highly likely at a certain time and place or context, and in such a scenario, the human ends up having to ask "manually" for the obvious.

Humans have gotten accustomed to the *Demand* model, which entails having to remember to perform various tasks such as monitoring their stock portfolio or keeping updated about new homes that have arrived on the market during a home search and home buying project. It can cause fatigue if the human keeps on aggressively demanding information during a period that the information has not changed (e.g., the stock portfolio value has been steady; and no new homes arrived on the market during this period of aggressive "demands"). On the other hand, the human can forget to demand information and miss out on being updated about changes in the portfolio value or new home arrivals on the market.

When the user is on the beach or hiking, for instance—in a *Demand* model—the user needs to remember to frequently "demand" information such as news for shark attacks, tides, crimes or violence, fires, or thunderstorms repeatedly and frequently, in order to keep abreast of real-world events pertinent to their current activity and context. Many or most of the demands will tend to return no significant new information, and this is a drawback of the *Demand* model.

2.4.2.2 Suggest Paradigm In the *Suggest paradigm*, the human allows the smart and intelligent agent-based automation to understand the context, analyze the user's history data as well as other relevant general data, and make reasonable inferences in order to proactively provide information or suggestions that have a high likelihood of being useful, timely, and relevant to the user. In the *Suggest* model, the user is notified automatically when the computer, driven by an intelligent agent-based system, has detected information and scenarios that justify a suggestion or notification or alarm to the user.

In case of the stock portfolio and home search examples earlier, the intelligent agent provides suggestions/notifications when something noteworthy and significant has occurred, such as a change in the stock portfolio beyond a threshold or when a new home has arrived on market that strongly matches the user's known search criteria.

2.4.2.3 Demand or Suggest? In the *Demand*-only model, the human on the beach or on a hike will need to repeatedly search the news manually for any recent shark attacks, thunderstorms, fires, and so on in order to keep abreast of such information. This can distract and detract from enjoying their real-world activities. In contrast to the *Demand* model, when the sophisticated *Suggest* model is in place, the user can relax and enjoy their real-world activity and be notified automatically when something noteworthy occurs. This can be accomplished in various scenarios such as by using the user's current location (i.e., Laguna Beach), the user's current context (on the beach with family), and automated scanning for current news and weather about the location, inferring the sentiment (such as danger), and then making the determination that an alert needs to be pushed out to the user. Such intelligent agent cloud-based computing scales well for a large set of users since the efforts of such computation is often performed, not for an individual user but for a set of users that are in the "same boat" or beach at a given point in time.

2.4.2.4 Demand and Suggest: A Healthy Balance *Demand* and *Suggest* are not mutually exclusive; rather, the new *Suggest* model ideally coexists with the *Demand* model. HCI has evolved over these years of mass adoption; it has matured to a point today that a healthy balance and mix of *Demand* and *Suggest* paradigms make an optimal interaction between human and computer possible. It saves time, is more efficient, and is based ideally on adaptive algorithms—whenever the user does not enthusiastically consume the routine suggestions of particular categories, back-off policies kick in, in order to make them less frequent.

2.4.3 Evolution of the Human–Computer Relationship

The human–computer relationship has evolved over time in the direction of "progressively deepening intimacy." For consumers, the computer itself had become smaller in size and weight, and its shape has become more elegant. From being located on a desk in front of the human—at arm's length—the computing devices moved closer by being perched on human laps and closer still by being housed in human pockets and purses or held in human hands for extended periods of time. The next step in the progression of this trend brings us to computing devices that are in contact with the human body for extended periods of time.

2.5 A Multi-Device World

Today, a consumer interacts with a wide variety of personal computing devices. There are often several computing devices per user in developed and developing nations. Such computing devices are located in varying degrees of proximity. Some devices may be worn on the body, such as a smart watch or sensor band, while other devices may be carried in the pocket or handbag. Some devices may be located in the home, some in the automobile, and so on.

2.5.1 Spatial Scope of Computing: Devices near and Devices far

In a world of a multitude of devices, networks, and services that a user interacts with, we find that there are some computers and networks that reside closer to the user and some progressively farther away from the user. Proximate networks include interconnected devices close to the user, while wider networks include a company or college campus, a citywide metro network, the Internet, and so on.

By organizing and categorizing the devices and networks in this manner and seeing computing from this perspective, it is easier to see how these various devices and networks can be made to work together and provide synergistic value.

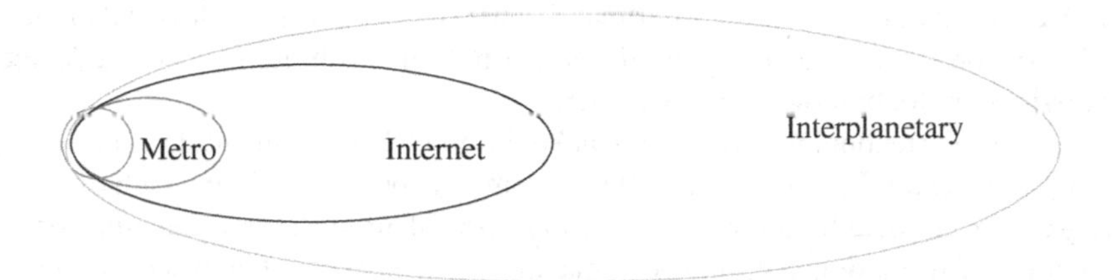

Figure 2-3 Spatial scope of computing—devices near, devices far.

In Figure 2-3, the smaller circles or ellipses represent devices and networks that are "proximate"—physically situated nearer to the user—such as the user's mobile phone, tablet, and smart watch. The ever-expanding larger ellipses represent computing resources such as cloud-based computing resources that reside farther away from the user, which the user nonetheless interacts with—via their proximate devices. Not all devices that are spatially closest to the user will necessarily connect to the Internet—which is why we find that the smallest ellipse is not wholly contained within the larger ones.

In such a nomenclature and categorization, the devices that the user may wear on their person are denoted as the body area network (BAN), while devices that reside within the user's home are denoted as the home area network (HAN).

2.5.2 Body Area Network (BAN)

A Body Area Network (BAN) is a wireless network of wearable computing devices that are centered around the human body. Such devices may be surface mounted or even embedded inside the body and are typically connected wirelessly over this BAN to a mobile smartphone or tablet device. The mobile devices can collect data from the body-worn devices and store such data; they may also, in turn, interconnect the BAN to the Internet. Thus, it is technically feasible for such body sensor data to be made available for remote monitoring by medical systems and so on.

Also, consumer fitness sensors and applications can help the user get insight into the various commonly understood metrics such as temperature, resting heart rate, average resting heart rate, and so on. Consumers can potentially share their fitness data with their doctors and health care providers. With the mass consumer usage of bodily fitness sensors in conjunction with cloud-based storage, there is potential for the power of large-scale computation to provide significant prediction, inferences, and forecasting. There is promise of a path to the much needed, affordable, and proactive health care—via use of Internet-based technology. Medical devices and applications are regulated by particular governmental agencies and need to comply with applicable laws, and these generally vary by country. Medical devices and applications are distinct from fitness sensor devices and applications, especially from a legal and regulatory perspective.

Some medical devices and applications merely monitor particular bodily parameters and are technologically quite similar to fitness devices and applications, but legally, they are distinct—fitness devices and applications are not regulated by the governmental agencies, that medical devices and applications are.

Other medical devices and applications regulate, actuate, or control some bodily parameters and functions and these are significantly distinct from fitness devices/applications, which do not control or actuate any bodily function. In this sub-arena, medical devices and fitness devices are technologically dissimilar.

The rate at which technology is evolving makes it somewhat difficult for laws and legislation to catch up or keep up. Legislation can often become the bottleneck in the path of innovation and progress in the health care arena. There is a lot of promise and potential for a more fundamental transformation in the delivery, management, and cost-effectiveness of health care that leverages the technological advances of lower-cost sensors, diagnostic software Apps that can run on generic lower-cost handheld (phone and tablet) devices, and beyond.

2.5.3 Personal Area Network (PAN)

The personal area network (PAN) is a network of interconnected devices that are centered around the user's living space; it includes devices that users carry with them including mobile devices such as smartphones, tablets, Chromebooks, netbooks, etc.; devices on the desk and devices in the home including home automation; and smart networked devices such as washers, dryers, refrigerators, and so on. The PAN can be considered to include the BAN, but it can also be relevant to think of the PAN as distinct from the BAN. In any case, the PAN and the BAN are spatially proximate and have opportunity for meaningful interconnection.

Bluetooth and IrDA are two of the common technologies that help interconnect the BAN and the PAN. IrDA has been around since the 1990s and is an industry standard and a set of protocols that address communication and data transfer over the "last one meter" by using infrared light. Bluetooth is a set of protocols that addresses wireless communication over distance of a few feet. Bluetooth Low Energy (LE), also known as Bluetooth Smart, aims to reduce the power consumption. A recent update to the Bluetooth specifications (version 4.2) addresses the direct connectivity of Bluetooth Smart devices to the Internet.

2.5.4 Home Area Network (HAN)

The Home Area Network (HAN) is a local area network (LAN) that interconnects devices that are within the home or within close proximity to the home. Such a network may include mobile devices and wired computers, televisions and entertainment devices, printers, scanners, thermostats, lamps, sprinklers, and so on. Most Internet service providers provide one IP address for the external network facing router. All the devices within the network typically have a local private IP address. The router that connects to the Internet service providers' network represents the boundary at which the Internet service provider's network ends and the home network begins. While the user is at home, the HAN may typically include the PAN and the BAN.

Network address translation (NAT) is a technique that hides the local IP address of a device behind a single device such as a router—which may have a public/external IP address. NAT is closely associated with IP masquerading—wherein one device such as a router masquerades as several other devices behind it—that have a local private IP address but appear as a single public/external IP address to the external public network. Such devices are able to initiate connections to the Internet, but are not directly addressable on the Internet. Most home networks use a NAT-based arrangement.

2.5.5 Automobile Network

There are devices and networks associated with our automobiles such as entertainment, locks, keys, and diagnostics. These form part of the automobile network and may overlap with the home network, while the automobile is parked within range of the home wireless network. The PAN and the BAN can become part of this automobile network when the user is within the vehicle.

2.5.5.1 Controller Area Network (CAN) There are formal protocols related to vehicles—a vehicle bus is a specialized internal communications network that interconnects components inside a vehicle (e.g., automobile, bus, train, industrial or agricultural vehicle, ship, or aircraft). Special requirements for vehicle control such as assurance of message delivery, nonconflicting messages, minimum time of delivery, and EMF noise resilience, as well as redundant routing and other characteristics, mandate the use of less common networking protocols. Protocols include controller area network (CAN), local interconnect network (LIN), and others. CAN bus is a vehicle bus standard designed to allow microcontrollers and devices to communicate with each other within a vehicle without a host computer. CAN bus is a message-based protocol, designed specifically for automotive applications but now also used in other areas such as aerospace, maritime, industrial automation, and medical equipment.

2.5.6 Near-Me Area Network (NAN)

A near-me area network (NAN) is a logical communication network that focuses on communication among wireless devices in close proximity. Unlike LANs, in which the devices are in the same network segment and share the same broadcast domain, the devices in a

NAN can belong to different proprietary network infrastructures (e.g., different mobile carriers). So, even though two devices are geographically close, the communication path between them might, in fact, traverse a long distance, going from a LAN, through the Internet, and to another LAN. NAN applications focus on two-way communications among people within a certain proximity to each other.

2.5.7 Campus Area Network

A campus area network is a network made of LANs that are interconnected within a geographical area and owned by a single entity such as a corporation or a university. Such a network typically has various relevant network services.

2.5.8 Metro Area Network

A metro area network is a network that spans a metropolitan area such as an entire city and is managed by a single, coordinating organization. Most networks are now aligning with the Ethernet-based metro Ethernet, which is used to connect subscribers to the larger networks including the Internet. There is tremendous potential to leverage metro area networks in many dimensions such as emergency management, community, and services. At a grand scale, the devices on the network and their activity and location provide a reflection of the current state, en masse of the human population, pets, resources, energy conservation, and so on. Such opportunities for efficiency and optimization of human activity, resources, safety, and so on can help realize the vision of the "smart city."

2.5.9 Wide Area Network

A wide area network networks include telecommunication networks that span national and international boundaries. The Internet, too, can be considered to be a wide area network.

2.5.10 Internet

The Internet is the global, interconnected network of networks that is based on the standard TCP/IP communication protocol. It consists of millions of public, government, academic, and business networks linked by a wide range of electronic, wireless, and optic fiber technologies.

There are two main name spaces in the Internet—the Internet Protocol (IP) address space and the Domain Name System (DNS) maintained by the Internet Corporation for Assigned Names and Numbers (ICANN). The technical standardization of the core IPV4 and IPV6 protocols are managed by the Internet Engineering Task Force (IETF), which is a nonprofit organization.

The modern Internet came into being sometime around the mid-1980s and initially was used predominantly in academic institutions. Commercialization occurred in the 1990s. However, the origins of the modern Internet date back to the 1960s and the research conducted by the US government as well as UK and France.

2.5.11 Interplanetary Network

Even though an interplanetary, galactic network seems a little like science fiction, an initial form of such a network already exists—the International Space Station is already connected to planet earth's Internet.

A wider interplanetary network requires a specialized set of protocols to address more of a store and forward approach that handles the delays and interruptions that could range from minutes to hours in view of the distances. One of such initiatives is the delay-tolerant networking (DTN), which is an architecture that endeavors to address technical issues in heterogeneous networks that lack continuous network connectivity including networks in space. At the core of DTN is the Bundle Protocol family—very similar to the Internet Protocol (IP)—which has been designed to account for the delays and disruptions expected in space communications.

2.6 Ubiquitous Computing

Ubiquitous computing is the computing paradigm of an always available access to computing resources in a coherent manner from any location and via one or more user-facing devices. The mobile era has set the stage for this model of "ubiquitous computing" to come into widespread practice, whereby data and computing is accessible from anywhere and at anytime—typically subject to network connectivity. Ubiquitous computing emphasizes universal access to computing as well as collaboration of devices over the network. Ubiquitous computing is known by other names such as "pervasive computing," "ambient computing," and so on.

2.7 Collective, Synergistic Computing Value

We have commenced, since several years, to interact with a wide variety and growing number of computing devices, via varying mechanisms of interaction. Our interaction with, and the value derived from computing devices, is (or ideally ought to be) less about interacting with one particular device and more about how these various devices might work together via interacting, interconnecting, and collaborating in order to assist us, save time and effort, and improve efficiency and productivity among many other such dimensions of our lives.

The computing environment, infrastructure, and our mindset has now matured and evolved to the point of being able to answer the questions of what these multitudes and groups of computing devices can do for us collectively and collaboratively, rather than what any particular device can do individually, in isolation.

Wearables can play a key role in such an ecosystem due to their proximity to the user and/or the physical environment.

2.7.1 Importance of the User Centricity and the User Context

User centricity in conjunction with a user context that transcends the existence or uptime of any particular device is particularly important and relevant. The transdevice user context is one of the key foundations of a more interactive, intelligent agent-based, ubiquitous model of computing.

As was the case decades ago, a given user interacted with about one personal device, that is, on a one-to-one basis such that device centricity happened to be mostly coincident with user centricity. But today, each user interacts with a multitude of devices, that is, on a one-to-many basis, so it becomes important to align with the user-centric model of data and context. A device-centric model tends to become obsolete in a world of many devices per user.

2.7.2 Distributed Intelligent Personal Assistant

An intelligent personal agent performs tasks, autonomously on an ongoing basis, in order to make human lives more convenient and safer. In a world of a multitude of diverse devices and in order to serve a user's needs at all times, the intelligent agent ideally runs not on any particular device, but as a distributed intelligent agent that runs across collaborating devices, which are user context aware at all times. The cloud is certainly the ideal candidate for the "headquarters" for such a distributed intelligent personal agent. The devices that reside close to the user also have great importance due to their ability to provide sensor signals that are the basis for the inferences about the user's current activity, context, and environment.

2.8 Bright and Cloudy: Cloud-based Intelligent Personal Agent

The foundation of the Suggest model is a sophisticated Intelligent Agent that is aware of the user's contexts at all times, watches out for the user's well-being at all times, and adapts its behavior based on learning algorithms. Such an intelligent agent on the cloud gives it a bird's-eye view and much depth and width of perspective.

Given a user who is driving and headed in a certain direction, the cloud-based intelligent agent can not only suggest alternate routes when the route ahead has traffic congestion, but it can also extrapolate all the various possible events such as ongoing car chases, ambulance paths, hurricanes, and so on that could affect the user's projected route—in order to make timely suggestions and recommendations to keep the user safer and on schedule for appointments and arrival destinations.

Analysis of the user's current location and context, in the backdrop of the news and events that occur moment to moment, is a full time job and one that can be best placed in the hands of this intelligent personal agent that resides primarily on the cloud, so the user is freed up to contemplate on or engage in matters of deeper significance.

The cloud provides both high-capacity storage and also tremendous processing power. The cloud-based systems can optimize computations for a group of users in the same "boat" or situation and benefit from the economy of scale. Cloud-based data centers are typically located close to sources of electric power, which reduces transmission losses and improves reliability.

2.8.1 Google / Cloud-Based Intelligent Personal Agent

Google, as the search engine, initially started out providing us access to information and knowledge. After having started out providing access to the world's information, Google has over the years become a prominent repository of the world's information. Over time, by virtue of mass usage and analysis of patterns and distribution in search terms, Google Insights acquired deep predictive capabilities. For many years, Google was able to predict an outbreak of the flu earlier and more accurately than the Centers for Disease Control and Prevention (CDC).

Google, as the cloud repository of the users' email, documents, photos, calendar, and location history stored on Google's secure cloud infrastructure, has the advantage of the best data and the best algorithms to create intelligent computing value for the consumers.

2.9 Leveraging Computer Vision

In the backdrop of IoT and an ubiquitous computing environment, there are sensors that can detect various real-world parameters of interest. These could include sensors embedded into the highway road's surface to detect traffic volume or sensors embedded in farmed land that detect moisture levels or moisture sensors embedded in the soil within the potted plant in your living room.

Video and audio sensors include cameras and microphones, which are often IoT devices, but they can include handheld and wearable devices. Video data from a section of highway can, for example, be analyzed in order to infer the traffic density without the need for sensors embedded into the highway road surface. The quality of the road surface can also be inferred by means of video data from a camera that covers a section of the road.

Computer vision is a field that includes the acquisition and processing of image data in order to infer relevant information. Computer vision aims to perceive and understand image data and, to some degree, duplicate the abilities of human vision. Computer vision algorithms can detect real-world objects such as human faces and cars in real time and thereby count cars and people that pass through a section of road or walkway. Some use cases of sensors embedded into surfaces for acquisition of particular real-world information can be addressed via image data and computer vision algorithms. Depending on the specifics, it may be less expensive and easier to install a camera and a vision-based system compared to a multiple specialized embedded sensors.

2.9.1 Enhanced Computer Vision / Subtle Change Amplification

While computer vision started out with the "modest" goal of duplicating human vision, researchers at the Computer Science and Artificial Intelligence Laboratory (CSAIL) at MIT, which include Professor William T. Freeman and Michael (Miki) Rubinstein (Ph.D.), have made progress in the arena of analysis and amplification of subtle motion and color changes.

Freeman and Rubinstein share with us the big world of small motions and changes in color, which when detected and amplified can give us "superhuman" vision. Rubinstein—as part of his Ph.D. research, with Freeman as his advisor—developed methods to extract and amplify subtle motion and color changes from videos. The beating heart results in the rhythmic flow of blood in the human body, which causes corresponding subtle, rhythmic changes in the color of the human skin. These color changes are generally invisible to human eye. Using the video signal from a regular camera and by detecting these subtle color changes via signal processing algorithms, it is possible to infer the heart rate. By creating a change amplified version of the video, it becomes easy to perceive the heart rate visually. This is an instance of vision enhancement via color change amplification. The human breath results in subtle elevation and movement of the chest and stomach area. Neonatal infants typically need to be monitored for vital signs while minimizing disturbing or touching them. The subtle motion of the belly after amplification yields a modified video which makes their breathing and the rate, thereof, visually obvious—without needing to touch or disturb them. Rubinstein also demonstrates other applications of this technology such as recreating a conversation by amplifying the movements of a crumpled bag of chips while a conversation is occurring in the vicinity. By zooming in on small motions of the crumpled bag and amplifying them in the order of a 100 times and after converting the motions into sound, the conversation can be constituted. A TED talk by Michael Rubinstein "A Big World of Small Motions" as well as other videos on this technology is available at:

https://www.youtube.com/watch?v=fenV3W7hQtw
https://www.youtube.com/watch?v=3rWycBEHn3s

Such amplification of subtle motion and color changes can be effectively applied to other arenas such as industrial, transportation, agricultural, and more. In one approach, a grid of moisture sensors embedded into the soil in a section of farmland provide useful input that can be used to time and control the periodicity of watering of the crop. In another approach, the video feed from a camera that "observes the crop" can be analyzed, and the subtle motions of the crop such as the swaying motion in the wind or any slight color changes or wilting can be amplified to detect commencement of crop dehydration and making it more visible and detectable for timing the watering accordingly.

Thus, the ordinary camera's video feed can be used to infer and reveal various measurements indirectly via sophisticated algorithms and computation.

2.10 IoT and Wearables: Unnatural and over the top?

So far in this chapter, we have covered the technological aspects of computing and networks. It certainly seems that a world with many devices, clouds, IoT, and wearables is unnatural and represents somewhat of an overdose of technology.

Just how unnatural and over the top is this world of IoT and wearables—smart devices, smart homes, sensors, intelligent agents, ubiquitous computing, and so on? The answer is subjective and depends on one's perspective. This section covers some correlations seen in nature and human history relating to networks and computation.

At a conceptual level, networks represent paths of transportation or flow of physical matter, services, and information. Tools represent extensions of biological intelligence and abilities. Computations involve sensing and recording data of data, analysis, and prediction. All these concepts are present in nature and not unique to human civilization or computer science. There is potential value that can be derived from the use of tools and computation, and there is the choice that one can make individually—to use or not use particular offerings that come from these technological advancements.

2.10.1 Human History of Tool Use and Computation

What we see today is, probably, merely an acceleration of a trend that was established and has been in place for tens or hundreds of thousands of years. Although technology has advanced in recent decades, there is no fundamental shift from our fundamental dependence on tools and computation to improve our daily lives and even our chances of survival.

Humans and their ancestors have been a tool maker and computational species. Long ago, tracking the movements of the sun, moon, planets, stars, and constellations helped our ancestors understand and predict celestial and astronomical cycles and the seasons; build clocks and calendars; and plan hunting, migration, and planting.

Observations of various patterns and associated predictions were the factor that enhanced the chances to human survival. Long ago, the rock chip gave human ancestors the competitive edge for survival. Tool use actually caused an increase in brain size and that bigger brain helped in building the next generation of more sophisticated tools.

The use of tools for hunting helped provide adequate meat more easily, which helped nourish the brain and improve its size and intelligence and provide more time to think and ponder and gaze at the skies and create rock art.

Not too much has changed, it would seem, because today it is the silicon "chip" that gives us the competitive edge, individually and as a species. It extends our memory, helps us consume, manage, create, and share information. It helps predict many aspects of human interest such as weather, finances, health, and more, to help us live more comfortably, safely, and efficiently.

Today, with the highest human population in recorded history inhabiting planet earth, perhaps the computing ecosystem of wearables, IoT, and artificial intelligence can solve many of the problems of infrastructure and resource management, energy efficiency, health monitoring, manufacturing efficiency, city and township management, and so much more.

2.10.2 Communication Networks in Nature

There are numerous examples of networks in nature that enable communication. It turns out that plants interconnect their roots with other plants—via the "mycelium," the branching threadlike, vegetative part of mushrooms (fungi) that grows in the soil—thereby forming a communication network via which information such as warning signals of pathogen and aphid attacks are transmitted between plants. Mycelium can be really tiny, and they can also be quite massive.

Mycelium is useful in nature and to the ecosystem in various roles—including the role of a communication network. Paul Stamets—the renowned mycologist and author of *Mycelium Running: How Mushrooms Can Help Save the World*—shares some of his insight into the world of mycelium, which form large 2000 plus acre networks in the forests of the Pacific northwest of the great North American continent. The mycelial network helps the overall health of the forest by distributing nutrition and information for the overall good of the forest. The mycelial network thrives in a healthy forest, and it strives to keep the forest healthy. Stamets points out in his writing and talks on TED (http://www.ted.com) that fungi are sentient beings that can sense the environment, human presence, and much more and that the mycelial network makes it possible for particular trees that do not receive adequate sunlight to receive nutrition via the mycelial network's ability to access and transport needed mineral nutrition.

Similar to the networks in nature, the important foundations of modern human civilization include the advanced systems for transportation, power transmission, water distribution, and information superhighway. Computation and communications have an important place in human civilization and are not necessarily all that artificial in concept. IoT and wearables have an important role of the sensor–actuator network, which addresses the sensing and detection, and feedback to the periphery of this network. The periphery of this computational network is what engages people and their relevant, significant "things" of interest directly.

2.10.3 Consumption of Power: by computational systems, biological and artificial

Our computational and artificial "intelligence" is built by, and is an extension of, our human biological brain. It turns out that the Internet somewhat resembles a massive organism and also our nervous system by demonstrating self-healing behavior, adaptation to changes in the flow of packets based on dynamic changes in available routes, redundant connections, and so on.

The smartphones and other user-interacting devices such as IoT and wearables lie on the periphery of this organism-like nervous system of the computational Internet.

The human brain comprises less than 3% of the total body weight, yet it accounts for over 18% of the energy consumption. Much like the human brain has this density of storage and processing, civilization's data centers have a density of storage and computational power as well as the need for massive amounts of energy to power this processing and data storage.

The subject of the power consumption by the cloud/data centers and the overall information technology industry has attracted much attention in recent years. It is estimated that if one were to count the energy consumption that goes into manufacturing the processors and devices, running the network infrastructure, and the data centers put together, then as much as 10% of the world's generated electricity is consumed in powering information technology.

It turns out that, much like the human brain has tremendous need for power, oxygen, and nutrition—compared to the rest of the body—the huge computational data centers, networks, and devices that run human civilization have huge needs of power and energy. The energy needs of the computational resources to run our human civilization have perhaps begun to mirror the relative power needs of the human brain with respect to the human body.

2.11 Security and Privacy Issues

In recent years, mass usage of mobile Apps and associated data collection on the cloud have brought the issues of security and privacy to the forefront of consumer attention. IoT and wearable devices raise the issues of security and privacy to an even greater degree, especially if IoT and wearable devices are directly present on the Internet as an independent device that has its own IP address and/or send data to cloud endpoints without the user's knowledge and control. It's more difficult for an isolated small IoT device with limited devices to possess an independent IP address and independent presence on the Internet and also defend and protect itself from malicious attack. As described earlier in this chapter, it is safer and more cost-effective for IoT and wearable devices to act as clients that maintain a secure possibly long-lived connection to cloud endpoints in a user-centric approach that gives the user control over their data. It is also easier for cloud-based endpoints to protect and defend themselves from malicious attack because of their access to computing resources, physical isolation, firewalls, algorithmic monitoring, and adaptive defenses.

Figure 2-4 shows the relationship between users, IoT devices, generated data, and cloud endpoints: a given user may own various IoT devices. IoT devices in turn may generate some form of recorded data and also send such data to various cloud endpoints on the Internet. IoT devices may also receive some data, instructions, or commands from the cloud endpoints.

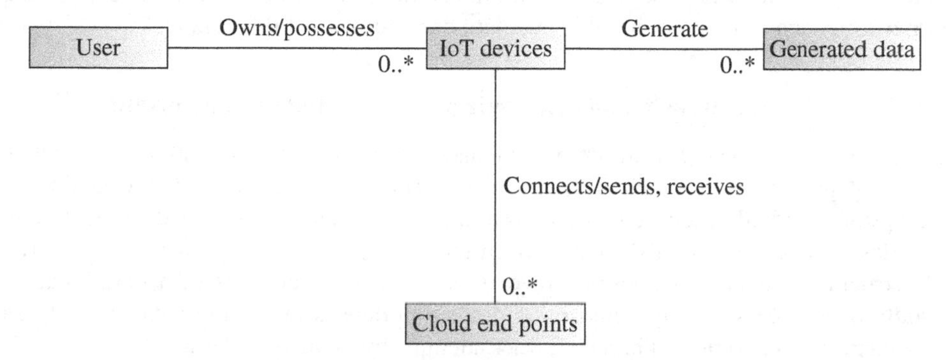

Figure 2-4 User, IoT devices, generated data, and so on.

In such a world of a multitude of IoT type smart devices, wherein the user's various devices may collect various data and send such data to various cloud end points such that the user is unaware of what data is collected and who the cloud-based entities are that store, analyze, and potentially share such data with their extended partners, is a scenario that has poor user privacy and control.

It does disservice to the users if their data is collected and stored but not accessible by the users themselves for their own needs and purposes. It is unfair to the user if they are not able to opt in or in, per their needs and choices.

In such a backdrop, the following section covers are some of the important principles that boost user privacy, security, and user control over data.

2.11.1 Use Awareness and complete end-to-end Transparency

It is important that users are aware of what data their IoT devices are generating, and to which cloud-based endpoints the data is being sent to, as applicable.

There have been many news reports in the media about household smart gadgets that record data and send it to the cloud without the user's knowledge and function like spyware.

Complete end-to-end transparency makes it very clear to the user what data is being collected and who the data is being sent to, how that data is being used, how securely the data is stored on the cloud, and if it is shared with external parties.

2.11.2 User Control and Choice

It is important that the users have control over the data collection with the ability to turn off the data collection per their needs and preferences—as well as the choice to opt in or opt out of data collection or sharing with external parties.

2.11.3 User Access to Collected Data and Erasure capability

It is important that the user has access to all of their data that is collected and stored on the cloud. It is also important that the user has the ability to erase the data stored on the cloud permanently and also export it out in standard open formats of their choice.

2.11.4 Device side, transit, and cloud side protection: Data Anonymization

It is important that the user is aware of what data is collected, who collects and stores it, the degree of protection that their data enjoys in terms of security on the device side, the encryption standards used, and the algorithmic protection the cloud side infrastructure provides in order to protect their data from unauthorized access and cyberattack. Strong infrastructure boundary protection prevents or reduces the chances of a data breach or unauthorized access. Strong algorithmic protection detects failed attempts promptly and intelligently in real time and blocks further attempts by malicious entities.

It's also important that the stored data be anonymized. Data anonymization is a technique of encryption and removal of personally identifiable information from sets of data. Anonymization maintains certain data centered around a random ID, rather than a identifiable individual. Sensitive data such as passwords, credit card numbers, and social security numbers need to be stored with strong encryption rather than as "plain text." Hashing algorithms such as MD5, which is often used for hashing passwords, is relatively easy to break. The National Institute of Standards and Technology (NIST) recommends PBKDF2 for one-way hashing.

2.11.5 Practical Considerations: User Centricity

It is impractical for users to have login account credentials on a per device and cloud/website basis—especially in a world of such a multitude of devices and cloud accounts. It is important that users use strong passwords that are not re-used across different realms. At the same time, it is important that users are able to securely access the history data collected on the Internet cloud endpoints and websites, control data access and sharing, and delete the data if they so desire. One of the solutions to address this problem is OpenID.

2.11.5.1 OpenID OpenID is an open standard and protocol that attempts to consolidate user's online identities so that users can log into various websites without having to register over and over. The OpenID-enabled website acts as the "relying party," which depends on the OpenID provider to authenticate their users. Users select accounts by first choosing an OpenID provider and the associated credentials.

OpenID providers include major Internet and technology companies such as Google, Yahoo, Facebook, Microsoft, WordPress, and several more. Such consolidation of the online identity is useful because the user can have far fewer login accounts with unique and more secure credentials and remember them more easily. OpenID 2.0 was finalized in December 2007 and OpenID adoption has been growing.

OpenID is decentralized—it does not rely on a central authority to authenticate a user. Furthermore, OpenID does not mandate any particular or specific set of authentication mechanisms—it can work as well with biometric authentication, smart card-based authentication, user name/passwords, and anything else in the future.

OpenID can be a win–win both for website owners as well as for users, because both can depend on the OpenID provider to address user authentication. Websites often find it challenging to maintain the user names and passwords and store them securely on their sites. Users find it harder to have to remember user names and strong distinct passwords for a large number of websites. It is important for the users' security that credentials at each site or IoT device be unique and complex.

2.12 Miscellaneous

A few miscellaneous topics are covered in this section.

2.12.1 PhoneBloks: Waste Reduction

PhoneBloks is a modular smartphone design concept created by Dutch Designer Dave Hakkens. "Bloks" are modular components that can be attached to the main board of the phone. Such "Bloks" can be upgraded and replaced, while retaining the rest of the device. Since many users end up replacing their entire phone every few years, there is a huge volume of electronic waste that is generated. By selectively upgrading a functional "Blok" such as camera, battery, storage, and so on, users can upgrade particular modules without giving up the entire device, thereby reducing electronic waste. PhoneBloks is an independent organization with the general mission of electronic waste reduction. More information on PhoneBloks can be found at https://phonebloks.com.

2.12.1.1 Project "Ara" Project "Ara" is an initiative from Motorola and Google that is influenced by, and with some degree of collaboration from, PhoneBloks. Project Ara aims to create modular smartphones and devices based on kits with modular components that can be put together like Lego blocks to create devices. Such modular components include common features such as camera and battery as well as specialized features such as game controller buttons, sensors, medical devices, receipt printers, laser pointers, and so on.

2.12.2 Google Cardboard: inexpensive Virtual Reality

Google Cardboard is an inexpensive cardboard headset designed by Google that works along with stereoscopic vision and display software on the Android smartphone—in order to provide users with 3-dimensional, virtual reality App experiences. Consumers can fold their own based on the designs provided by Google or purchase a ready-made version from various manufacturers listed on their site. The prices start at about US $15. More information can be found at https://www.google.com/get/cardboard. Google also has an associated virtual reality Cardboard SDK that simplifies the development of virtual reality Apps, the coverage of which is outside of the scope of this book.

References and Further Reading

http://en.wikipedia.org/wiki/Wearable_computer#History

http://en.wikipedia.org/wiki/Mechanical_computer

http://www.media.mit.edu/wearables/

http://en.wikipedia.org/wiki/Casio_Databank

http://en.wikipedia.org/wiki/Calculator_watch

http://en.wikipedia.org/wiki/Nelsonic_Industries#Game_Watches

http://www.media.mit.edu/wearables/lizzy/timeline.html

http://www.cs.virginia.edu/~evans/thorp.pdf

http://www.amazon.com/Beat-Dealer-Winning-Strategy-Twenty-One/dp/0394703103

http://en.wikipedia.org/wiki/Alex_Pentland

http://web.media.mit.edu/~sandy/

http://www.forbes.com/forbes/2010/0830/e-gang-mit-sandy-pentland-darpa-sociometers-mining-reality.html

http://en.wikipedia.org/wiki/Punched_card

http://en.wikipedia.org/wiki/Punched_card_input/output

http://en.wikipedia.org/wiki/Mechanical_computer

http://en.wikipedia.org/wiki/Infrared_Data_Association

http://en.wikipedia.org/wiki/ARM_architecture

http://en.wikipedia.org/wiki/MIPS_Technologies

http://en.wikipedia.org/wiki/List_of_system-on-a-chip_suppliers

http://en.wikipedia.org/wiki/Internet_of_Things

https://hbr.org/2014/10/the-sectors-where-the-internet-of-things-really-matters

http://en.wikipedia.org/wiki/Machine_to_machine

http://en.wikipedia.org/wiki/Computer_vision

http://en.wikipedia.org/wiki/Microelectromechanical_systems

http://en.wikipedia.org/wiki/Metro_Ethernet

http://en.wikipedia.org/wiki/Delay-tolerant_networking

http://en.wikipedia.org/wiki/Network_address_translation

https://www.google.com/cloudprint/learn/howitworks.html

http://www.gartner.com/newsroom/id/2636073
http://nfc-forum.org
http://en.wikipedia.org/wiki/Metro_Ethernet
http://en.wikipedia.org/wiki/Delay-tolerant_networking
http://en.wikipedia.org/wiki/Microelectromechanical_systems
http://www.ncbi.nlm.nih.gov/pubmed/24129903
https://phonebloks.com
https://www.google.com/get/cardboard/

Part II Foundation Android

This part covers the Android operating system (OS) and Android development platform from the ground up. Although Android applications are written using the Java programming language, the runtime environment that Android applications execute in is not a Java runtime. Although the Android OS is based on the "Linux" OS, Android is not a standard Linux "distribution" — there are a number of modifications that are unique to the Android kernel or core of the OS. Gaining a little insight into Android's underlying relationship with Linux and Java, as well as incorporating some of the vocabulary of its underlying building blocks into our technical dictionary, will go a long way toward understanding Android's core concepts better. Such awareness and understanding can be useful in debugging and troubleshooting an Android application, when the root cause of a particular issue reveals itself from deeper down in the stack.

This section also provides a basic overview of the Android SDK as a foundation for covering the *Android Wear* and *Google Fit* platforms. *Android Wear* and *Google Fit* represent particular subsets of the Android platform. An *Android Wear* application is an Android application that is targeted to a particular API level, namely, API level 20, which represents the Wear platform. A *Google Fit* application is an Android application that uses the *Google Fit* (Fitness) API. In case you already have experience with Android application development on Android 5.0/Lollipop, you may skip particular topics in this part of the book. In case you are relatively new to Android development, this part provides an overview of the Android SDK from the ground up and lists resources for further reading. However, it is possible that you may need more resources to complement this book's coverage of the basic Android platform. This section also covers the topic of interdevice communication, which is relevant to a world with a multitude of devices and peripherals.

Wearable Android™: Android Wear & Google Fit App Development, First Edition. Sanjay M. Mishra.
© 2015 John Wiley & Sons, Inc. Published 2015 by John Wiley & Sons, Inc.

Chapter 3 Android Fundamentals / Hello Lollipop

3.1 Android: Introduction

The *Android operating system* (OS), which powers the majority of the smartphones sold in the world today, is a relatively new phenomenon. The Android platform was unveiled in 2007 along with the founding of the *Open Handset Alliance*™ (OHA)—a consortium of hardware, software, and telecommunications companies. Most of the major semiconductor manufacturers and handset makers are part of the OHA. The first Android phone was sold in 2008.

Android can be said to have originated in 2003 at **Android Inc.**, a startup cofounded by **Andy Rubin,** then reportedly working on a mobile OS. The phonetic correlation between "Android" and "**Andy Rubin**" is most likely not a coincidence. Google acquired the 22-month-old Android Inc. in 2005. Andy Rubin led the development on the Android platform for about a decade, taking it from nothing to an impressive position, in a relatively short span of time. Android Inc. continues to be a separate company that is owned by Google Inc. This separation between Google and Android extends into the SDKs and API namespaces.

Although Android is a relatively new platform, Android's origins go way back in time and have a close connection with the Linux/Unix family of OS as well as the widely used *Java* programming language—which are technologies that have been around since several decades. The Android OS is derived from Linux, a free and "Open Source" OS.

Wearable Android™*: Android Wear & Google Fit App Development*, First Edition. Sanjay M. Mishra.
© 2015 John Wiley & Sons, Inc. Published 2015 by John Wiley & Sons, Inc.

3.2 Linux: "*nix" or Unix-like OS

Linux, also known as GNU/Linux, is a Unix-like or "*nix*" OS. The *nix family of OS has historically powered the majority of servers on the Internet. Today, the *nix family of OS dominates the world of supercomputers as well as mobile devices and small embedded devices including Wearable and IoT devices. The Unix family of OS has a long and complex history, as detailed below.

3.2.1 Unix

Unix is a multiuser, multiprocessing OS, with strong networking and security. The Unix OS dates back to the late 1960s: Ken Thompson and Dennis Ritchie at the AT&T Bell Labs conceived of and commenced implementing the UNIX® OS around the year 1969. Unix was first released around 1971. Unix had been written initially using assembly language but was rewritten using the C programming language (for the most part) in 1973. The use of a higher level language "C" made Unix more portable to diverse platforms and processor architectures. Due to some antitrust issues at that time, AT&T was forbidden from entering the computer business and was required to license the Unix source code to anyone who asked for it. Unix commenced to be used in universities and businesses since the 1970s and became more widely used since 1980. By the mid-1980s, AT&T divested itself from Bell Labs, and Bell Labs was free from any legal obligations regarding selling UNIX® as a

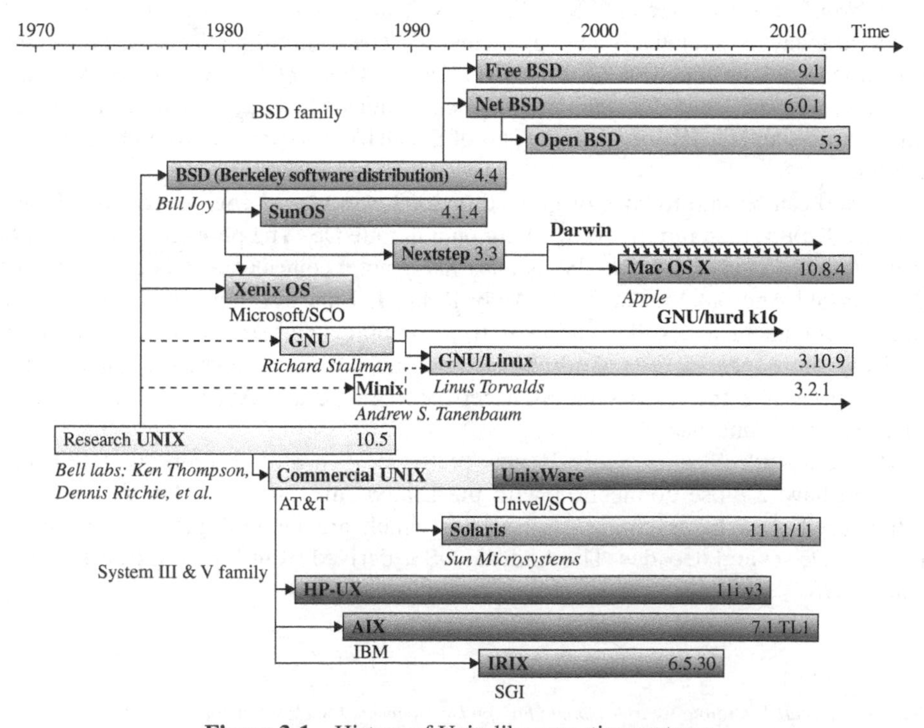

Figure 3-1 History of Unix-like operating systems.

proprietary product. Since the 1980s and into the 1990s, a multitude of proprietary flavors of Unix were available from Digital, HP, Sun Microsystems, and IBM—along with the corresponding computer hardware. UNIX System V was the commercial version of Unix first releases in 1983, which had four major revisions. System V R4 was the major effort by the major vendors to collaborate on and unify Unix. Today, various Unix System V OS continue to be available from IBM, HP, and Oracle® and are currently in use in businesses and academic institutions, in data centers worldwide.

Figure 3-1 shows a simplified representation of the history of the Unix and Unix-like, *nix family of OS, which happens to include Mac OS X®, Solaris®, HP-UX®, and AIX®.

Though not indicated in the figure, both Apple's iOS® and Android are also part of the Unix-like family of OS. This speaks to the success and longevity of the Unix-like family of OS, which has evolved and adapted over the decades in its myriad of flavors, forms, and brands, on diverse chipsets and device form factors. The device form factor refers to various aspects of the devices' physical characteristics, such as screen size and density, camera, hard buttons, internal components and layout, and more. In the case of Android, we tend to find a large and wide-ranging set of device form factors in the market place—due to the open nature of the platform. Historically, the *nix family OS has supported diversity of chipsets and server/device form factors.

3.2.2 Open Source

The term Open Source means that the source code is available to users under a license in which the copyright holder provides others the right to study, modify, and even distribute the software—compliant with certain licensing terms and conditions. Open Source software projects are often developed collaboratively, openly, and transparently. The Open Source model is different from the "proprietary" model where the source code is under a restrictive copyright and typically unavailable to the users of the software. The Open Source software tends to spur innovation and can result in cost savings to consumers and businesses. The Linux OS and the Apache Web server are two classic examples of highly successful Open Source projects. The Android OS is another such example. The Open Source nature of the Android OS makes the public release of the vast majority of its source code inevitable. It also enables the downstream use and modification of Android source code, consistent with the applicable license terms.

3.2.3 GNU / Free Software Foundation

In the mid-1980s, Richard Stallman started the GNU project and the Free Software Foundation (FSF). The FSF is a non-profit organization that supports the development of free software. The GNU project is a set of free software development tools and compilers—such as *make*, *gcc*, *gdb*, and *glibc* (an implementation of the standard C library)—that have been developed and made publicly available by the FSF. GNU tools are based on Unix-like design concepts. GNU is a recursive acronym for "gnu's not unix"—while UNIX® was a proprietary OS associated with expensive hardware in the 1980s, GNU's original aims were to create a free Unix-like OS for the relatively inexpensive and highly popular Intel 8086 family of processors. GNU software is Open Source and covered by the GNU General Public License (GPL) or its variants. The Open Source Linux kernel is covered by the GPL.

Today, the Linux OS and the Android OS are built using the GNU toolchain. *make* and *gcc* are some of the core programs in the GNU toolchain, and these are used to build these OS. The box below shows some basic output from these programs and the copyright notice of the FSF.

```
~ $ make --version
  GNU Make 3.81
  Copyright (C) 2006 Free SoftwareFoundation, Inc.
  This is free software; see the source for copying conditions.
  There is NO warranty; not even for MERCHANTABILITY or FITNESS
  FOR A
  PARTICULAR PURPOSE.
  This program built for x86_64-pc-linux-gnu
~ $ gcc --version
  gcc (Ubuntu/Linaro 4.8.1-10ubuntu9) 4.8.1
  Copyright (C) 2013 Free Software Foundation, Inc.
  This is free software; see the source for copying conditions. There is NO
warranty; not even for MERCHANTABILITY or FITNESS FOR A PARTICULAR
PURPOSE.
```

While the kernel—as the name suggests—is the very core of the OS, a Linux distribution is a more usable form of the OS that includes a collection of useful software bundled around the Linux kernel. There are over 600 Linux distributions, though only a handful are widely known. Community-driven Linux distributions include Debian, Slackware, and Gentoo, while company-backed distributions include Ubuntu, Fedora, and OpenSUSE.

3.2.3.1 Free as in Freedom: GNU Public License

According to the principles of the FSF:

> "A program is free software if the program's users have the four essential freedoms:
>
> The freedom to run the program, for any purpose (freedom 0).
>
> The freedom to study how the program works, and change it so it does your computing as you wish (freedom 1). Access to the source code is a precondition for this.
>
> The freedom to redistribute copies so you can help your neighbor (freedom 2).
>
> The freedom to distribute copies of your modified versions to others (freedom 3). By doing this you can give the whole community a chance to benefit from your changes. Access to the source code is a precondition for this."

Freedom to distribute (freedoms 2 and 3) means you are free to redistribute copies, either with or without modifications, either gratis or charging a fee for distribution. The only condition is compliance with the GNU GPL, which mandates that you continue to make the code with your modifications available under the GNU GPL and to make transparent and explicit any changes that you may have made. The Linux OS kernel is covered by and released under the GNU GPL. Due to the transitive nature of the GPL, the Android kernel, which derives from the Linux kernel, is also covered by the GPL. This is what makes the Android kernel "Open Source" and covered under a GPL.

3.2.4 Apache Software Foundation: Apache Software License

The Apache License is also an Open Source software license—from the Apache Software Foundation. It requires preservation of the notice and disclaimer, but it does not require derivative works or modifications of the software to be distributed using the same license. The Android Open Source Project (AOSP) commenced by Google, which includes the Android frameworks and nonkernel portions of Android, is also Open Source but covered by the Apache Software License.

3.3 Linux: yesterday and today

In the late 1980s, Linus Torvalds, then a young Finnish student, commenced on creating a free Unix-like OS for the highly successful and economically priced Intel 80386 processor of that era. Linux was built using the free GNU complier and tools such as *gcc* from the FSF. Torvalds released Linux in the early 1990s under the GNU license, and Linux was born. Torvalds initiative took off from where the Stallman's GNU tools and compliers left off. By the mid- to end 1990s, Linux evolved, matured, and was ported successfully to run on many diverse processor families, with relative ease due to its modular and flexible design that underwent massive refinement over the years. Today, Linux is a highly successful and thriving OS that runs on most widely used processors/chipsets on supercomputers and servers that run the most used websites, and also consumer and embedded devices such as routers, printers, smartphones, tablets, IoT, and Wearables.

Torvalds owns the "Linux" trademark and manages the Linux kernel's code contributions, reviews, and merges to this day. The TIME® magazine places Linus Torvalds among the top 100 most influential people in the world, which includes physicist Albert Einstein, molecular biologists James Watson and Francis Crick, computer scientist Alan Turing, moviemaker Steven Spielberg, and singer/songwriter Bob Dylan. Torvalds was placed 17th in the "TIME 100: The Most Important People of the Century poll" held in 2000. There is at least one asteroid named after Linus, namely asteroid "*9793 Torvalds*."

3.4 Unix System Architecture

Unix is a powerful, multiprocessing, and multiuser OS characterized by features that include built-in networking (TCP/IP), persistent system services called "daemons," and files modeled as abstractions for devices and objects.

3.4.1 Unix Processes

A process is a program in execution—a program that is currently executing. A program itself consists of object code stored on disk or any type of media. A process is much more than merely the program's object code and includes additionally a data section, a set of resources such as open files and pending signals, an address space, and one or more threads of execution. Threads of execution, or simply threads, are objects of activity. Linux has a unique implementation of threads such that the distinction between a thread and a process is somewhat indistinguishable at a very low level. Each process has an associated process identifier (PID) and a parent process identifier (PPID).

3.4.1.1 Linux Processes The **ps** command is one of the standard commands available on most *nix systems including Android, and reports a snapshot of the process status, as the name *ps* suggests. The following is a partial listing of the output of the **ps** command on my Ubuntu Linux development host. We see later about 11 of the 250 odd processes running on my Ubuntu Linux development host (Figure 3-2A).

```
~ $ ps -eaf | head -12
UID        PID  PPID  C STIME TTY          TIME CMD
root         1     0  0 18:11 ?        00:00:01 /sbin/init
root         2     0  0 18:11 ?        00:00:00 [kthreadd]
root         3     2  0 18:11 ?        00:00:00 [ksoftirqd/0]
root         5     2  0 18:11 ?        00:00:00 [kworker/0:0H]
root         7     2  0 18:11 ?        00:00:00 [migration/0]
root         8     2  0 18:11 ?        00:00:00 [rcu_bh]
root         9     2  0 18:11 ?        00:00:00 [rcuob/0]
root        10     2  0 18:11 ?        00:00:00 [rcuob/1]
root        11     2  0 18:11 ?        00:00:00 [rcuob/2]
root        12     2  0 18:11 ?        00:00:00 [rcuob/3]
root        13     2  0 18:11 ?        00:00:00 [rcuob/4]
~ $ █
```

Figure 3-2A Linux processes on Ubuntu.

3.4.1.2 Android Processes Figure 3-2B shows a partial listing of running Android processes, as seen from the output of the **ps** command. About 15 of the 200 processes running on a Nexus 7 with the Android version 5 (Lollipop) OS are seen in the diagram.

Each process is associated with a user who "owns" the process, a PID, a PPID, and the name of the program that's executing in that process.

As we will be seeing later in this book and in great detail, in the Android world, whenever an App is started, a process needs to be assigned or started to house and execute the App's object code and components. This process is owned typically by a user ID that is distinct for every App. Each App runs in its own sandbox and is subject to its security context and permissions model, imposed by the Android OS. Apps that are signed by the same key are associated with the same OS user ID—as we will cover in more detail shortly.

3.4.1.3 Process Tree Linux processes are organized in a hierarchy. Each process has a PPID, except for process 0. The processes at the root of this process are the scheduler

```
/opt $ adb shell ps | head -15
USER     PID   PPID  VSIZE  RSS     WCHAN    PC         NAME
root      1     0     744    504    ffffffff 00000000 S /init
root      2     0     0      0      ffffffff 00000000 S kthreadd
root      3     2     0      0      ffffffff 00000000 S ksoftirqd/0
root      6     2     0      0      ffffffff 00000000 S migration/0
root      16    2     0      0      ffffffff 00000000 S khelper
root      17    2     0      0      ffffffff 00000000 S suspend_sys_syn
root      18    2     0      0      ffffffff 00000000 S suspend
root      23    2     0      0      ffffffff 00000000 S irq/203-msmdata
root      24    2     0      0      ffffffff 00000000 S sync_supers
root      25    2     0      0      ffffffff 00000000 S bdi-default
root      26    2     0      0      ffffffff 00000000 S kblockd
root      27    2     0      0      ffffffff 00000000 S msm_slim_ctrl_r
root      28    2     0      0      ffffffff 00000000 S khubd
root      29    2     0      0      ffffffff 00000000 S irq/84-msm_iomm
/opt $ █
```

Figure 3-2B Android processes on Nexus 7, Android 5 (Lollipop).

(sched), init, and kthreadd, and their process IDs typically are respectively 0, 1, and 2. The scheduler process (PID 0) is the parent of the init process (PID 1). All the various processes originate from the init process (PID 1) and ultimately from the scheduler process (PID 0).

Figure 3-3 shows a partial listing of a Linux process tree on Ubuntu. The init process (PID 1) is the parent of the other processes such as cron, NetworkManager, and bluetoothd, to name a few.

```
~ $ pstree | head -21
init-+-NetworkManager-+-dhclient
     |                |-dnsmasq
     |                `-3*[{NetworkManager}]
     |-accounts-daemon---2*[{accounts-daemon}]
     |-acpid
     |-avahi-daemon---avahi-daemon
     |-bluetoothd
     |-colord---2*[{colord}]
     |-console-kit-dae---63*[{console-kit-dae}]
     |-cron
     |-cups-browsed
     |-cupsd
     |-2*[dbus-daemon]
     |-dbus-launch
     |-6*[getty]
     |-irqbalance
     |-lightdm-+-Xorg---2*[{Xorg}]
     |         |-lightdm-+-init-+-adb---3*[{adb}]
     |         |         |      |-at-spi-bus-laun-+-dbus-daemon
     |         |         |      |                 `-3*[{at-spi-bus-laun}]
     |         |         |      |-at-spi2-registr---{at-spi2-registr}
```

Figure 3-3 A partial listing of a Linux process tree.

3.4.1.4 Unix Interprocess Communication (IPC) Interprocess communication (IPC) refers to the communication and data exchange between different processes. There are many mechanisms that help and support IPC. These include pipes, signals, semaphores, message queues, shared memory, and sockets.

3.4.1.5 Remote Procedure Calls (RPC) A remote procedure call (RPC) is a high-level mechanism for IPC and distributed client–server communication, in which the calling program and the called procedure exist in separate address spaces. The two processes may reside on the same system (host) or they may be on different systems. RPC is also known as remote invocation, especially when the software has been written in an object-oriented programming language. RPC isolates and abstracts out the transport protocols from the application logic and makes it easier to write applications. Android has its own distinct and lightweight mechanism for IPC, as we will cover shortly.

3.4.2 Unix Kernel

The Unix kernel is the core of the OS and consists of the key subsystems for process management, memory management, device management, network management, concurrency and multitasking, scheduling, and so on. The kernel also provides concurrency, interrupt handling, separation of user space from kernel space, system calls, file descriptor management, and more (Figure 3-4).

User space	Applications Libraries
Kernel space	Process management Memory management Device management
Hardware	CPU Memory Devices

Figure 3-4 Unix kernel.

The Unix kernel is intentionally kept small and focused on its core function. Actual kernel implementations can be based on a monolithic design or a micro/modular design. A monolithic kernel is a single executable that runs as a process. A microkernel (also known as modular kernel) entails multiple binaries and multiple processes that carry out the responsibilities of the kernel.

3.4.2.1 Linux Kernel The Linux kernel is based on the Unix design principles and is based on a monolithic design. The Linux kernel is the basis for a large number of Linux family "*nix" OS. The kernel is covered by the GPL V2 license.

3.5 Java

Java is one of the most popular programming languages today and has millions of developers. One of its key characteristics is the "write once, run anywhere" approach—developers can write and compile a program once and run it on other OS platforms. Several aspects and features of the Java platform have conceptual equivalents in the Android platform.

3.5.1 Java Origins

Java began as a Sun Microsystems internal project around 1990, headed by James Gosling who is widely known as the "father of Java." Java was originally aimed at embedded devices and set top boxes for televisions. Soon, Java morphed and evolved into a desktop as well as an Internet and Web programming platform. By the late 1990s, Java had become a very popular and widely used programming language, which emphasized portable threading, security, memory management, and garbage collection.

3.5.2 Java Platform: Language, JVM

One of the Java platform's key characteristics is that there are well-defined specifications for the language and the runtime and a separation of specification from the implementation. There are several Java platforms—Standard Edition, Enterprise Edition, Micro Edition, Java Card, Embedded, Real time, and more. The Java Standard Edition is the base Java platform, and there are a multitude of implementations from various vendors.

The Java Language Specification specifies the Java language, and these specifications can be found at http://docs.oracle.com/javase/specs. The Java® virtual machine (JVM) refers to the runtime code execution component and subset of the Java platform. A JVM instance is a particular implementation's running process that executes Java bytecode. The JVM has a multitude of implementations. The list of JVMs is available at http://en.wikipedia.org/wiki/List_of_Java_virtual_machines and shows a multitude of such implementations. The JVM includes the garbage collector, which addresses the automated deallocation of memory.

3.5.3 Java memory: Heap, Stack, and native

The heap is the region of memory where Java objects reside, and it is shared between all threads within the JVM process. Heap memory may increase or decrease in size while the program executes. Garbage collection helps free up unused objects occupying the heap. The JVM parameters *-Xms* and *-Xmx* help you specify the starting and maximum heap sizes as shown in the snippet below:

```
$ java -Xms512m -Xmx1024m MyProgram
```

In the absence of an explicit setting for the heap size, the JVM uses certain default values that depend on the system that the program is being run, as well as other JVM flags, such as client or server mode settings.

Each thread has its own private stack memory, created at the same time that the thread is created. The default stack size per thread is in the order of around 256K or 512K. The stack size can be set using the JVM parameter *-Xss*. The JVM stack, much like in languages like C, holds local variables and partial results associated with method invocation and returns.

Native memory is the memory that is used by the *java* process; the *java* process itself is typically a native, non-Java-based program.

3.5.4 Security Policy: Permissions

The Java platform defines permissions that can be granted to a Java application via the *java.policy* file. The default *java.policy* file resides in the *lib/security* directory under the JRE installation home. Applications can be executed using custom *java.policy* files, and some of the permissions include *java.net.NetPermission*, *java.net.SocketPermission, java.security.AllPermission*, and so on. The permission AllPermission is special in that it includes and covers everything—as in all the other available permissions such as *java.net.NetPermission*, *java.net.SocketPermission*, and the rest.

In theory, *AllPermission* should be granted with care. In practice, many Java applications that are written and used in the real world do not leverage the fine-grained individual permissions and use *java.security.AllPermission* instead. The following snippet shows the *java.policy* file:

```
$ head $JAVA_HOME/jre/lib/security/java.policy
// Standard extensions get all permissions by default
grant codeBase "file:${{java.ext.dirs}}/*" {
        permission java.security.AllPermission;
};
// default permissions granted to all domains
....
```

The Android platform, as we'll be covering shortly, uses an elaborate permission-based mechanism, which is conceptually similar to the Java permissions model; however, it has no equivalent of *AllPermission*—which means that Android enforces fine-grained security. More information on Java security policy and permissions can be found at:

http://docs.oracle.com/javase/7/docs/technotes/guides/security/PolicyFiles.html
http://docs.oracle.com/javase/7/docs/technotes/guides/security/permissions.html
http://docs.oracle.com/javase/tutorial/security/userperm/policy.html

3.6 Apache Harmony

Apache Harmony is a free Open Source Java implementation under Apache License, commenced around 2005/2006 with significant backing from major companies led by IBM. Apache Harmony implemented incomplete, near-complete approximately 98% implementations of Java versions 5 and 6. The project was abandoned in 2010 after Oracle acquired Sun Microsystems, and Oracle and IBM joined hands to collaborate on the OpenJDK project. Android's App container environment uses the java class libraries from the free Apache Harmony Java, thereby distancing its runtime container's java namespace support from the Java implementation from Sun/Oracle.

3.7 Android OS and platform

The Android OS comprises of the Android kernel, which has versions for the processor family such as ARM, MIPS, and x86. Android kernel flavors vary by manufacturer and device. The manufacturers include Asus, Motorola, LG, Samsung, Texas Instruments, and so on. The models have code names that differ from the brand names that users see.

The list of kernels is available at source.android.com

https://source.android.com/source/building-kernels.html#figuring-out-which-kernel-to-build.

When you shell into your Android device, you will likely see a prompt that tells you the code name of your device.

```
shell@flo:/ $
```

Referencing the code name from the listing in the link above will tell you that the *flo* device has a Qualcomm MSM processor. Referencing /proc/cpuinfo on my device yields the Qualcomm processor model, as expected:

```
Hardware: QCT APQ8064 FLO
```

3.7.1 Android Kernel

Android is a Linux-based OS with a forked, monolithic kernel that supports the ARM, MIPS, and x86 processor families. The Android OS aims to provide a processor agnostic, secure application execution environment. Much like Java popularized the concept of "write once, run anywhere," Android too provides a hardware platform agnostic application development ecosystem.

The Android kernel has been modified to address the unique needs and constraints of mobile devices such as limited resources (slower smaller CPUs, smaller RAM, lack of swap space, and battery powered) and intermittent network connectivity. In addition, the Android OS provided an application execution container environment that executes processor-independent Dalvik Executable (DEX) bytecode suitable for running on the Android Runtime (ART). Android uses the bionic C library as its standard C library, rather than the GNU C library (glibc). Google developed the bionic C library, as a derivation of Berkeley Software Distribution's (BSD) standard C library to cater to the smaller memory footprint and optimizations for lower-frequency CPUs.

Unlike other typical OS including other Linux distributions, in the case of Android, the user and owner of the Android device do have root access. The Android OS and the */system* partition are read only. Although based on the Linux kernel, it has a large list of modifications and it excludes the GNU C library, which is one of the core components found in Linux distributions.

Thus, Android may be considered to be a Linux distribution or may be not—because expert opinions are divided. While the Android kernel is derived from GNU Linux, it has some unique customizations for mobile devices with limited resources and differing capabilities. Such a fork is not ideal, and it is likely that at some point in the next few years the Android kernel changes will get merged back into the Linux kernel.

Some of the Android kernel changes, incorporated additions from third-party Open Source projects and internal initiatives, and customizations that are unique to the Android kernel are listed below, along with a brief description:

Yet Another Flash File System (YAFFS) (incorporated)

Mobile phones typically and predominantly use flash memory and a flash file system is a file system for storing and retrieving files on flash memory. YAFFS was developed by Charles Manning for the company Aleph One and released under GPL. The Android kernel incorporates YAFFS as an addition with respect to the Linux kernel.

Binder (incorporated)

Android does not use the GNU/Linux IPC mechanisms and uses instead the lighter weight, Binder mechanism from the OpenBinder project. Binder serves as the low-level mechanism that supports communication between application processes and Android's system services as well as other application processes. Android has a high-level Android Interface Definition Language (AIDL), conceptually similar to Java Remote Method Invocation (RMI), which is supported by Binder at the low level.

Android shared memory ("ashmem") (added)

Ashmem is the shared memory allocator that allows processes to share the RAM across processes. One process may create a region of shared memory and share the corresponding file descriptor with another process. Ashmem is a low-level feature that some system server components depend on.

Android Logger (added)

The Android Logger provides system logging and support for Android's *logcat* command. The *cat* command on *nix systems concatenates files and prints them to standard output, which by default is your terminal. *logcat* is a command available on the Android device and can be executed when you access a shell or terminal in the device over *adb*. The following snippet shows the *logcat* command invoked directly on the Android device:

```
shell@flo:/ $ ls -l /system/bin/logcat
-rwxr-xr-x root     shell      17752 2014-10-15 20:47 logcat
shell@flo:/ $ logcat
--------- beginning of system
I/Vold    ( 177): Vold 2.1 (the revenge) firing up
I/InstallerConnection( 541): disconnecting...
I/SystemServer( 541): Entered the Android system server!
I/SystemServiceManager( 541): Starting com.android.server.pm.Installer
```

The *adb* command can be invoked on your development machine with various options and subcommands. Particularly, *adb logcat* or *adb shell logcat* has the effect of giving you access to the connected Android device's logs from your development machine. The following snippet shows the output from the *logcat* command, available on your development machine while connected to the Android device with debugging enabled:

```
$ adb logcat | head -5
--------- beginning of system
I/Vold ( 177): Vold 2.1 (the revenge) firing up
I/InstallerConnection( 541): disconnecting...
I/SystemServer( 541): Entered the Android system server!
I/SystemServiceManager( 541): Starting com.android.server.pm.Installer
$
$ adb shell logcat | head -5
--------- beginning of system
I/Vold ( 177): Vold 2.1 (the revenge) firing up
I/InstallerConnection( 541): disconnecting...
I/SystemServer( 541): Entered the Android system server!
I/SystemServiceManager( 541): Starting com.android.server.pm.Installer
$
```

Wake locks (added)

The Android system may partially turn off the screen or the CPU when the device is not actively in use, in order to conserve power. Wake locks are a mechanism that allows applications to indicate to the Android system that they need the device to stay on in order to perform some application tasks.

Android alarms (added)

Android has an AlarmManager as one of its system services, which the kernel supports at the lower level.

Paranoid network security (added)

Android restricts access to network functionality, unless the requesting user ID belongs to particular groups. This mechanism supports specifying application-level permissions that govern access to the network, Bluetooth, and so on.

Android Debug Bridge ("adb") (added)

The Android Debug Bridge (*adb*) is a command line tool that is used for debugging. It has three components—a daemon (*adbd*), which runs in the background of the Android device; a server (also *adbd*); and a client program *adb*, which runs on the development machine.

Figure 3-5 shows two *adbd* processes, one running on my Android device and the other running on my development machine. These two respective *adbd* processes connect and

```
$ adb shell ps | grep adbd
shell     200   1     8760   208   ffffffff 00000000 S /sbin/adbd
$ ps -eaf | grep adbd
sanjay    4954  4526  0 08:37 pts/0    00:00:00 grep --color=auto adbd
$ which adb
/opt/androidsdk/platform-tools/adb
$ which jdb
/opt/jdk1.7/bin/jdb
$ which gdb
/usr/bin/gdb
$ ■
```

Figure 3-5 adbd and adb.

maintain a debug bridge between your connected Android device and your development device. The *adbd* daemon that runs on the Android device is started only if the device debugging has been enabled via the Android OS Settings. When the *adb* command is run on your development machine, it checks to see if the *adbd* process is already running on your development machine, and in case it is not, it starts up the *adbd* daemon.

adb is in the same genre of debugging tools such as *gdb* (the GNU debugger) and *jdb* (the Java debugger that is part of the Java JDK), which have been in existence since decades in the world of Linux and Java. Therefore, the figure also shows as background information, the *jdb* command available on your development machine as part of the Java JDK installation. Although *jdb* is not widely used directly, it is often leverages under the covers by development and debugging tools and Java IDEs. The figure also shows the *gdb* command, which is a part of the *GNU* tools and happens to be installed on my development machine.

3.7.2 Android Open Source Project (AOSP)

The Android Open Source Project "AOSP" consists of the nonkernel part of the Android OS source code that includes the Android framework. The Android application framework addresses the App container environment that includes the *java.** namespace packages from Apache Harmony as well as the *android.**

While the Android kernel—the core of the Android OS—is covered predominantly by the GPL, the AOSP—non-kernel—portion of the source code is covered predominantly by the Apache Software License.

3.7.2.1 Android Framework The Android framework consists of the container environment and infrastructure for Apps to be installed and run, securely and reliably. It includes the various system services that Apps can avail of and provides the implementation for the *android.**, *java.**, and *dalvik.** namespaces.

3.7.3 Android Development

Android App development emphasizes the use of the Java programming language and recommends the use of the Java SDK from Sun/Oracle. The Android build process however converts the compiled java class files into "dex" files.

3.7.3.1 Android SDK The Android SDK is a Java-based SDK and represents the primary development platform for Android App development. The use of Java in the development environment and the platform independent *ART* on the Android device makes your App independent of the processor architecture of the hardware devices on which your App will run. One of the advantages of developing Apps for Android is that your App can usually be built once to run on a diversity of devices from various manufacturers, on diverse processors. As an App developer, you are generally insulated from having to deal with whether the device on which your App is running uses an ARM or x86 or MIPS-based processor. The only exception to this is if your App needs to depend on a native C/NDK-based library built by you or from some third party.

As a side note, there is an overloading of the term "native" that has crept into common usage and terminology—you may come across the usage of the term "native" to mean "native Android" App (built using the Android SDK) versus a cross-platform HTML5-based application that aims to provide a single solution across different mobile and desktop platforms. Based on the context, you will generally be able to distinguish the intended meaning of a given instance of usage of this term.

While on the subject of cross-platform Web applications that can address desktop and mobile platforms, generally, the more complex an App, the stronger the case for a platform-specific App and its associated, engaging user experience. You will find that most prominent Internet sites and brands struggled with this question and eventually created mobile platform-specific Apps that are meant to be downloaded and installed. For implementing some trivial functionality, perhaps a mobile-friendly Web application might suffice—with less of a justification for implementing an App.

With much of the business logic implemented as cloud-based API, the platform-specific client model tends to provide better responsiveness and a more elegant user experience.

3.7.3.2 Android NDK The Android NDK represents a secondary and niche development kit that is based on the ***C*** programming language. Your expertise in and/or preference for the ***C*** programming language does not make a sound case for using the Android NDK for App development. Opting to use the Android NDK on an arbitrary basis will tend to make most Apps unnecessarily complicated and introduce potential problems without any performance benefits. If your App needs to implement some CPU-intensive processing such as signal processing, custom encoding and decoding, and so on, that may make a case for leveraging the Android NDK to build a native library or write a native App.

More information on the Android NDK is available at https://developer.android.com/tools/sdk/ndk/index.html.

We will not be covering the installation of the NDK in this book; the references and further reading section at the end of this chapter provides more information for interested readers.

3.7.4 Android Runtime Environment

In contrast to the Android development time environment, the ART is Dalvik based and executes DEX code rather than Java bytecode.

This difference between the two environments can introduce complexities—in case you include libraries in your application from the namespace *org.json* or *org.bouncycastle*, for instance, which happen to be part of the Android stack, and there may be some effects from this. At development time, your code may compile, but during runtime, you may in some corner case scenarios encounter issues due to differing versions of these libraries, in resolving some methods that are not available or have differing signatures from their counterpart versions from the Android OS stack. The new Android build system and *Android Studio* address these matters elegantly.

3.7.4.1 Dalvik Virtual Machine The Dalvik virtual machine™ is the process and virtual machine (VM) that runs Apps. Dalvik is a part of the Android OS. Dalvik is internally a register-based VM (vs. the stack-based Java VM). Dalvik has been optimized for resource-constrained hardware such as phones. Dalvik's unique optimizations include a register-based instruction set and faster interpreter speed and the ability of the device to run multiple instances of the VM efficiently. Dalvik was written from the ground up and is a "clean-room" implementation, per Google. A "clean room" is an environment and design effort that excludes proprietary knowledge of a competitor.

3.7.4.2 ART (Android Runtime) *Dalvik* was the original runtime environment—the part of the OS that executes the application container-based Apps. ART is the new Android runtime that was introduced experimentally in the Android KitKat (4.4) release. In Android KitKat, users could switch the runtime between Dalvik and ART, via Android KitKat's *Developer Options* in *Settings*. In Android version 5 (Lollipop), ART is the default runtime. ART offers better application performance, superior garbage collection, and improved profiling and diagnostics.

3.7.4.3 Zygote In biology, zygote is the initial cell from which more cells divide and form. The Android OS runs each App in its own separate VM, within a separate OS process owned by a distinct OS user ID—for reasons of isolation and security. The actual user ID does not matter to the App; the actual user ID is used by the OS to enforce security-related functions such as keeping each App's files and data private.

In order to reduce the overhead of "cold" starting a separate VM for each App, Android has some optimizations in place—the *zygote* process is initialized at boot time and maintains a live VM instance with preloaded and initialized core libraries. *zygote*'s parent is the *init* process. *zygote* forks new VM instances on demand. *zygote* is a process and live VM instance that is initialized during the boot process. The read-only core Android libraries are shared across instances and not duplicated per VM instance, which reduces the memory needs of each VM. The *zygote* process thus speeds up the start time of Apps.

The box below shows the zygote process listing on an Android device. You can get a similar listing by running the command adb shell ps | grep zygote on your development computer after you have set up your Android development environment:

USER	PID	PPID	VSIZE	RSS	WCHAN	PC	NAME
root	195	1	1473464	54596	fffffff	00000000 S	zygote

3.7.4.4 System Server: Android System Services The first process that is forked off zygote is the *system_server* process (Figure 3-6A). That's because even before any App can be run, the various system services need to be initialized so they can support the needs of Apps.

```
$ adb shell ps | grep init
root      1     0     2616    756    ffffffff 00000000 S /init
$ adb shell ps | grep zygote
root      195   1     1473464 54620 ffffffff 00000000 S zygote
$ adb shell ps | grep system_server
system    534   195   1687516 129036 ffffffff 00000000 S system_server
$ adb shell
shell@flo:/ $
```

Figure 3-6A init, zygote, system_server processes.

The *system_server* process includes a family of "system" services that Apps avail of by calling the *getSystemService()* method available in the *Context* class. These system services include the *ActivityManager*, *PackageManager*, *AccountManager*, *NotificationManager*, *PowerManager*, *WiFiManager*, and so on. The Android system server is a part of the Android framework (Figure 3-6B).

```
$ adb shell cat  /proc/534/task/*/comm | grep -i manager
PowerManagerSer
ActivityManager
PowerManagerSer
PackageManager
AlarmManager
WifiManager
$ adb shell cat  /proc/534/task/*/comm | grep -i net
NetworkMonitorN
NetdConnector
NetworkStats
NetworkPolicy
EthernetService
NetworkTimeUpda
$ adb shell cat  /proc/534/task/*/comm | grep -i wifi
WifiP2pService
WifiStateMachin
WifiService
WifiWatchdogSta
WifiManager
WifiScanningSer
WifiRttService
WifiMonitor
$
```

Figure 3-6B Listing of live system_server threads, system services.

3.7.5 Android Interface Definition Language (AIDL)

Android Interface Definition Language (AIDL) is an interprocess mechanism, in the same vein as Corba IDL and Java RMI. AIDL much like other IDLs helps define the interface that clients can communicate with service components using IPC. The Android platform addresses the marshaling and unmarshaling that is involved in such communication. AIDL supports several data types off the shelf, which include the primitive types in the *Java* programming language, as well as Strings, Lists, and Maps. Applications can define their own custom data types, and these will need to leverage the *Parcelable* interface. The *android.os.Parcelable* interface is conceptually analogous to the *java.io.Serializable* interface used for object serialization in Java. Binder is the low-level mechanism that supports AIDL.

More details on AIDL, *Parcelable*, and Binder can be found at:

http://developer.android.com/guide/components/aidl.html

http://developer.android.com/reference/android/os/Parcelable.html

http://developer.android.com/reference/android/os/Binder.html

3.8 Setting up your Android Development Environment

Now that we have covered some of Android's background and system-level details, let us proceed with the setting up of an Android development environment. The Android development environment is far from monolithic—fundamentally, it uses a combination of tools provided by a multitude of vendors and entities.

There are four high-level artifacts that make up your Android development environment: *Java JDK*, *Android SDK*, build tools (*gradle* and *ant*), and the IDE (*Android Studio*). It is important that you set them up carefully and make a note of the directories where you install them. Details of the *Android SDK* and App development are available at https://developer.android.com.

The *Android SDK* is available for *Linux*, Mac®, and Windows® OS. Currently, it requires *Java JDK* version 7. Please be guided by the instructions and system requirements for your specific OS, per the details available at

http://developer.android.com/sdk/index.html#Requirements
as well as
http://developer.android.com/sdk/installing/index.html?pkg=tools.

In case you are using *Ubuntu*, for versions 13.x and above, you will need to install several packages using the following three commands: *sudo dpkg --add-architecture i386, sudo apt-get update*, and *sudo apt-get install libncurses5:i386 libstdc++6:i386 zlib1g:i386*. For *Ubuntu* versions 12.x and earlier, you will need to install one package via the command *sudo apt-get install ia32-libs*. This has been covered in the section "Troubleshooting Ubuntu" in the "Tool" documentation links provided above. However, in case you see any references to using *sudo apt-get install* for purposes of installing the *Java JDK*, please ignore them. The explanation for this is available in the next section.

3.8.1 Installing Java SDK version 7 (JDK 1.7) from Sun Microsystems / Oracle

As a Java developer, you have likely installed numerous versions of the Java SDK as well as various Java IDEs. Typically, Java developers install the "stand-alone" Java SDK (also known as the JDK), which comprises all the basic Java development tools such as *javac*, *javap*, *javah,* and *java*—the JVM. It's also ideal to set a *JAVA_HOME* environment variable set to the location of the JAVA_SDK installation and to adjust the *PATH* variable to include $JAVA_HOME/bin at the very beginning of the PATH. Making these settings in the *~/.profile* or similar file will make these settings consistently available every time.

```
export JAVA_HOME=/opt/jdk1.7
export PATH=$JAVA_HOME/bin:$PATH
```

After this, the command *which java* should show you the full path of the java executable.

```
~ $ which java
/opt/jdk1.7/bin/java
~ $
```

The Java JDK is available for download from:

http://www.oracle.com/technetwork/java/javase/downloads/index.html
http://www.oracle.com/technetwork/java/javase/config-417990.html

Since Java 7 is not a current release from Oracle (Java 8 is, at the time of writing), you will find Java 7 listed under "Previous Releases." Depending on your OS and whether it is 32 bit or 64 bit, you will need to download the appropriate flavor of the Java 7 (JDK 1.7) binary.

On Linux systems, using the command uname -mpio will help you determine whether your OS is 32 bit or 64 bit. The command man uname provides you with information about the various uname options (Figure 3-7).

The output of the *uname* command indicates that my development machine has an x86_64 OS, which is why I chose the *Linux x64* binary (jdk-7u67-linux-x64.tar.gz).

```
$ uname -mpio
x86_64 x86_64 x86_64 GNU/Linux
$ uname -a
Linux acer-ubuntu13 3.11.0-14-generic #21-Ubuntu SMP Tue Nov 12 17:04:55 U
TC 2013 x86_64 x86_64 x86_64 GNU/Linux
$ ▌
```

Figure 3-7 uname, determining your OS specifics.

In case you are using Ubuntu, I recommend not using the *sudo apt-get install*-based approach for installing the Java SDK. Similarly, I recommend not using the rpm format in case you are using a Linux flavor that works with rpms such as Fedora—for reasons explained in Section 3.11.

As we saw earlier in this chapter, there are a multitude of Java JDK providers such as IcedTea, GNU Classpath (GCJ), and many more. The right Java SDK for Android development needs to come from the source listed in the Android developer documentation. Once you have your Java SDK installation in place, the next step is to install the Android SDK tools.

3.8.2 Installing Android SDK from Google

The *Android SDK* is distinct from the *Android Studio* IDE, much like the *Java JDK* is distinct from any *Java IDE*. In this section, I will cover the installation of the "stand-alone" *Android SDK* first, followed by the installation of *Android Studio* in a subsequent step. *Android Studio* is the official and new IDE for Android development. *Android Studio* is based on the *JetBrains® IntelliJ IDEA®*, which is an excellent IDE for Java development.

Installing and maintaining a stand-alone *Android SDK* will help identify what tools and functions it provides. This approach helps in understanding the Android SDK better and can make it easier to troubleshoot if something does not work right. It also gives you more flexibility with using an alternative IDE for Android development. The difference between the two approaches is marginal. In either case, the end result is that your local environment will have the *Android SDK* and the *Android Studio* IDE working together, hand in glove. If you feel strongly inclined toward using the "cobundled" *Android Studio* and *Android SDK*, by all means go for it—Section 3.8.4 covers *Android Studio*.

When you visit the Android SDK home page (http://developer.android.com/sdk/index.html), you will notice a link to "**Other Download Options**" under the prominent *Android Studio* graphic (Figure 3-8A).

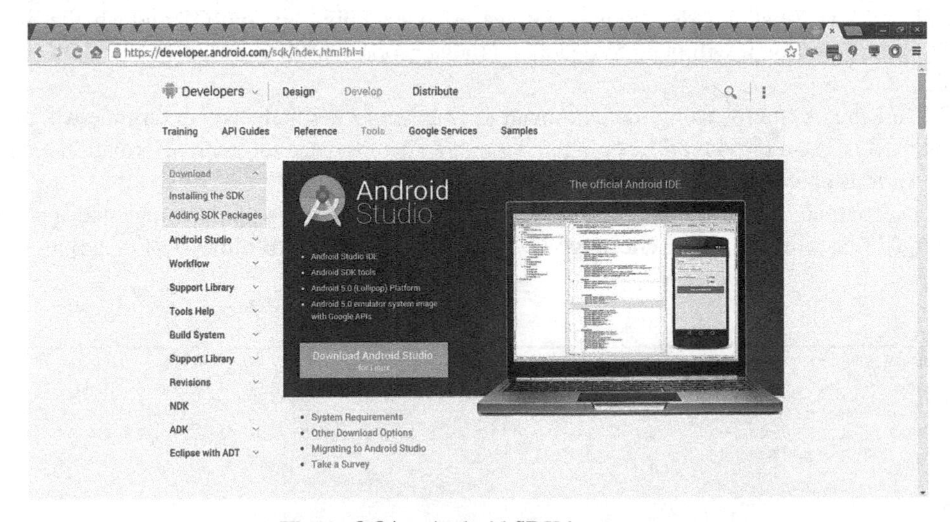

Figure 3-8A Android SDK home page.

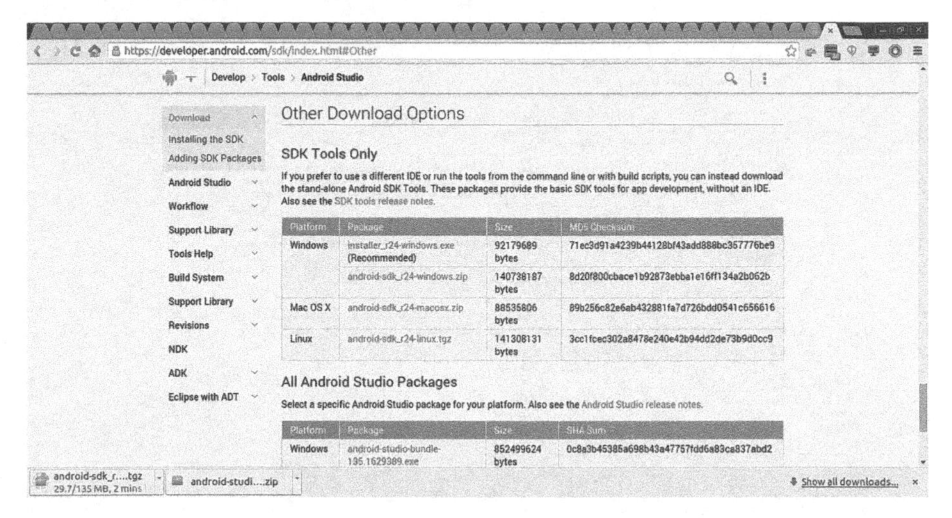

Figure 3-8B Android SDK other download options.

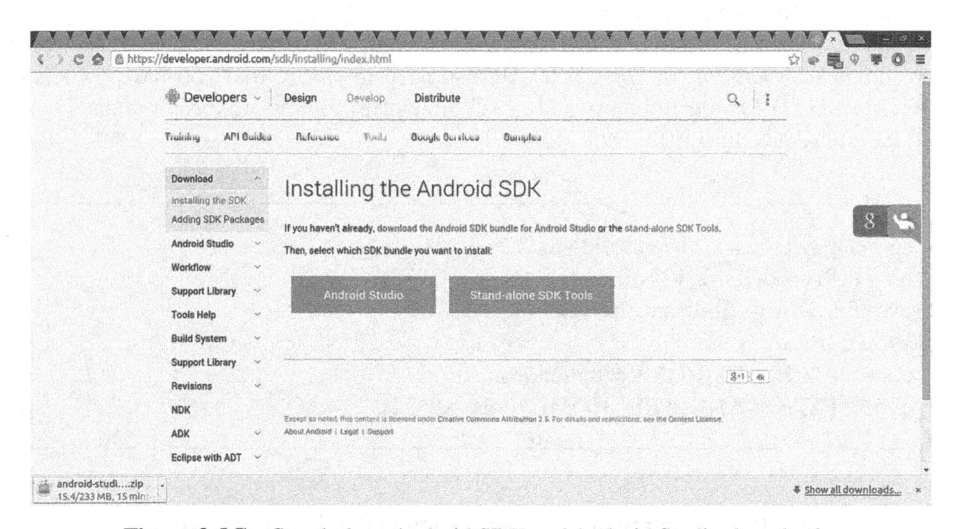

Figure 3-8C Stand-alone Android SDK and Android Studio downloads.

Clicking on *Other Download Options* takes you to a page that displays the "stand-alone" *Android SDK tools* and *Android Studio* as two separate binaries (Figures 3-8B and 3-8C).

In order to proceed with the download of the stand-alone *Android SDK*, you will need to accept and acknowledge the license agreement for the *Android SDK* (Figure 3-8D).

Once you have downloaded both the stand-alone *Android SDK* ("SDK Tools only") and *Android Studio* separately, you will need to proceed with extracting them in your local environment.

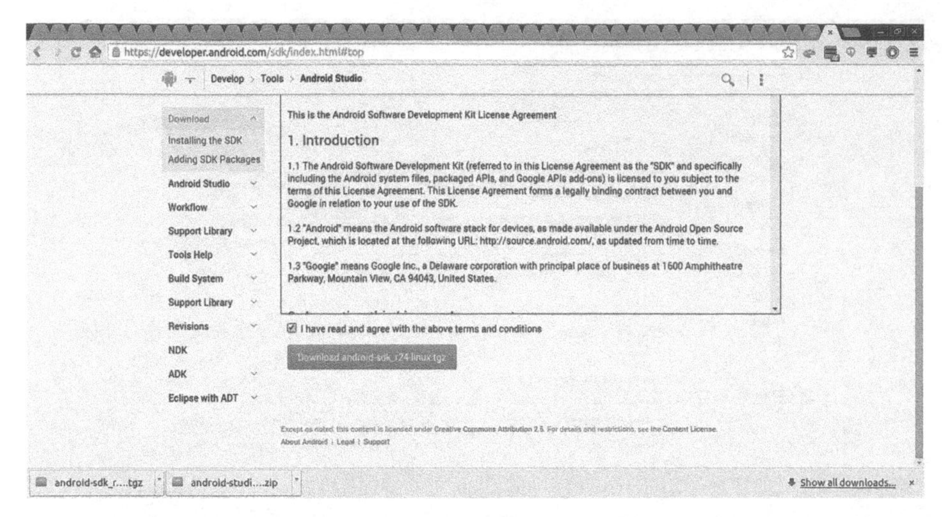

Figure 3-8D Stand-alone Android SDK tools license acknowledgment.

Extracting the Android SDK binary

Once the download of the Android SDK has completed, you can extract it to a location or your choice such as */opt* or *~/opt*. You will also need to ensure that you have ownership of the *Android SDK Home* directory and its subdirectories. Below are the commands I used as an indicative reference.

```
cp ~sanjay/Downloads/android-sdk_r24-linux.tgz .
tar zxvf android-sdk_r24-linux.tgz
mv android-sdk-linux androidsdk
chown -R sanjay
export ANDROID_HOME=/opt/androidsdk
export PATH=$ANDROID_HOME/tools :$PATH
```

I find it more effective to rename the installation home directory to something simpler such as *androidsdk*, rather than the original *android-sdk_r24-linux*. A simpler name is easier to validate, remember, and type in anytime as needed. I maintain a text file with the content "*android-sdk_r24-linux*" for future reference.

At this point, you should have the following environment variables in your shell environment along the lines:

```
ANDROID_HOME=/opt/androidsdk
JAVA_HOME=/opt/jdk1.7
PATH=$JAVA_HOME/bin:$ANDROID_HOME/tools :$PATH
```

Your shell environment should have both *java* and *android* from their respective, correct installation locations.

Updating your Android SDK

The extracted *Android SDK* directory tree initially contains a minimal set of content and tools. In order to be useful, it needs to be updated first. Therefore, the next step is to update your Android SDK. Typing android on the command line will present you with the *Android SDK Manager* screen.

The *Android SDK Manager* helps manage your local *Android SDK* installation and keep it updated by downloading packages and artifacts from the official cloud-based repositories. It shows you the state of your local environment with respect to what is available in the cloud repos, and helps you install and delete packages such as the various *tools*, *API* levels, and *extras* on your local environment. Each Android version has an associated API level. For instance, Android 5 (Lollipop) has the corresponding API level of 21 (Figure 3-8E).

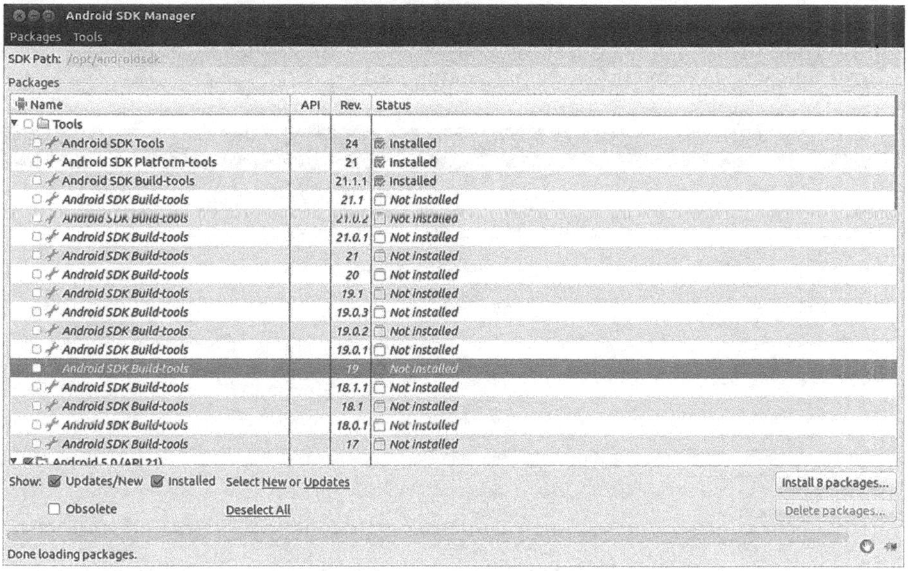

Figure 3-8E Android SDK Manager, SDK update.

In case your development machine is in a network that has a proxy server, you will need to enter in the proxy server's details under *Tools → Options*.

You will find that the *Android SDK Manager* has three logical sections under packages:

[***tools***]

[Various ***API*** levels (such as API 21 or Android 5 and so on)]

[***extras*** (such as *Google Play Services, Android Support Library, Google USB Driver,* ...)]

You will also find various revisions of the ***tools*** from 17 through 24 and beyond. It would be best to select all of them for installation.

Similarly, you will find various **Android API versions** such as API 21 (Android 5) or higher toward the top, down to API 3 (Android 1.5) at the very bottom. The set that you select will depend on what versions of Android you intend your Apps to support. If you are interested in installing a minimal set of packages, then selecting the build and platform tools Rev 20 onward along with the API level 21 (Android 5/Lollipop) and API level 20 (Android Wear 4.4W.2) is an essential choice for this book. If you are planning on writing new Apps targeted at recent versions of Android, a reasonable selection set would be API 15 (Android 4.0.3) through API 21 (Android 5) and beyond. If you are expecting to support Apps on Android 2.3.3 (API 10), you should ensure that it is selected.

With regard to ***extras***, it would be best to select them all in one shot.

You may also decide to select "everything" in all the sections—this will take longer to download and take up more disk space on your development machine, but other than that it can't hurt (Figure 3-8F). Over time, you will find yourself compiling and building a diversity of projects, libraries, and sample code, so having a wider range of versions of tools and platforms upfront can come in handy in the long run.

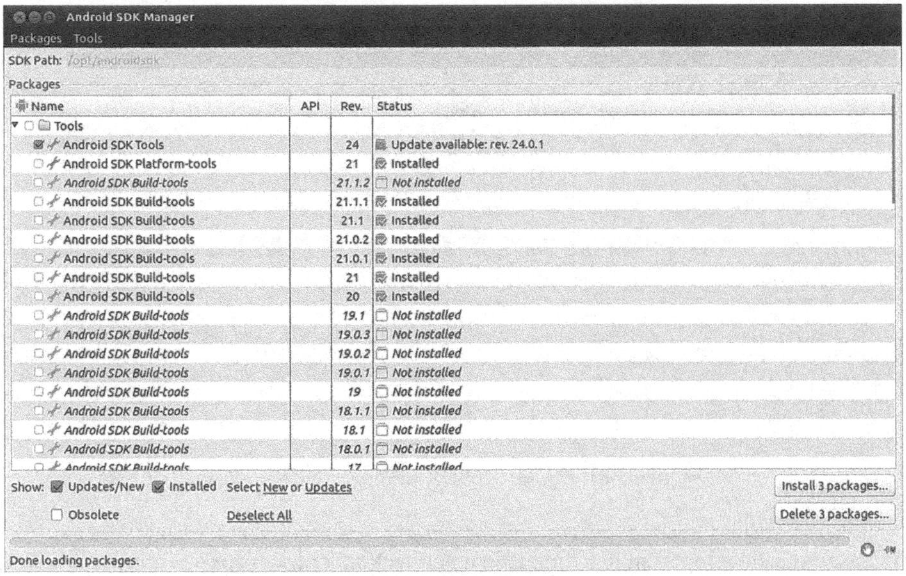

Figure 3-8F Selecting packages for installation.

Once you have selected the set of packages, you may click on the *Install Packages* button that is located toward the bottom of the screen.

You will find that there are various licenses involved that you will need to accept each of them individually before proceeding further (Figure 3-8G). After the download completes (Figure 3-8H)—which can take a while depending on the speed of your Internet connection and the number of packages you have selected—it would be a good idea to exit the SDK manager and repeat the android command, just to be sure you have downloaded

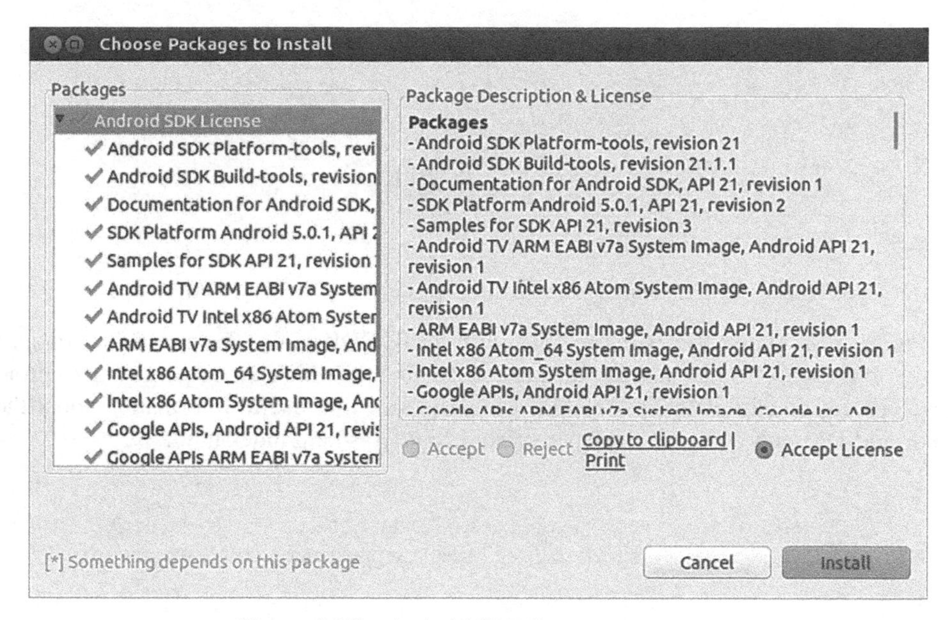

Figure 3-8G Android SDK license acceptance.

all of the intended packages. Sometimes you may have missed out on accepting a particular license, which will cause some of the packages to be left out, in the first pass.

You might have noted references to *Android SDK tools*, Android *platform tools*, and Android *build tools* in the Android SDK Manager or otherwise. There happen to be separate corresponding subdirectories under your Android SDK installation directory.

Android SDK Manager

Packages Tools

SDK Path: /opt/androidsdk

Packages

Name	API	Rev.	Status
ARM EABI v7a System Image	21	1	Installed
Intel x86 Atom_64 System Image	21	1	Installed
Intel x86 Atom System Image	21	1	Installed
Google APIs	21	1	Installed
Google APIs ARM EABI v7a System Image	21	3	Installed
Google APIs Intel x86 Atom_64 System Im	21	3	Installed
Google APIs Intel x86 Atom System Image	21	3	Installed
Sources for Android SDK	21	1	Installed
Android 4.4W.2 (API 20)			
SDK Platform	20	2	Installed
Samples for SDK	20	3	Installed
Android Wear ARM EABI v7a System Imag	20	4	Installed
Android Wear Intel x86 Atom System Ima	20	4	Installed
Sources for Android SDK	20	1	Installed
Android 4.4.2 (API 19)			
SDK Platform	19	4	Installed
Samples for SDK	19	6	Installed
ARM EABI v7a System Image	19	2	Installed
Intel x86 Atom System Image	19	2	Installed

Show: Updates/New Installed Select New or Updates Install packages...

Obsolete Deselect All Delete packages...

Done loading packages.

Figure 3-8H Android SDK—done loading packages.

```
$ ls
android          emulator64-mips   lib             screenshot2
ant              emulator64-x86    lint            source.properties
apps             emulator-arm      mksdcard        support
ddms             emulator-mips     monitor         templates
draw9patch       emulator-x86      monkeyrunner    traceview
emulator         hierarchyviewer   NOTICE.txt      uiautomatorviewer
emulator64-arm   jobb              proguard
$
```

Figure 3-9A Listing of Android SDK tools in $ANDROID_HOME/tools.

Figure 3-9A shows the listing of the core Android SDK tools. The core SDK contains the essential set of tools such as the *android* command, which are required for Android development on any platform and API level. You will find useful information about the Android SDK tools at http://developer.android.com/tools/help/index.html.

```
$ ls
adb  dmtracedump  fastboot   NOTICE.txt         sqlite3
api  etc1tool     hprof-conv source.properties  systrace
$
```

Figure 3-9B Android SDK "platform" tools in $ANDROID_HOME/platform-tools.

Figure 3-9B shows the listing of the "platform" tools. The *adb* command is the most commonly used platform tool.

The next category of SDK tools is the "build" tools, which are specific to the Android version/API level.

```
$ ls
17.0.0  18.1.0  19.0.0  19.0.2  19.1.0  21.0.0  21.0.2  21.1.1
18.0.1  18.1.1  19.0.1  19.0.3  20.0.0  21.0.1  21.1.0  21.1.2
$ pwd
/opt/androidsdk/build-tools
$
```

Figure 3-9C Listing of Android SDK build tool versions.

Figure 3-9C shows the various versions of the build tools installed on my development machine. Every time you install or upgrade an Android platform version/API level such as 19 or 21, these tools are potentially updated.

```
$ ls
aapt                          jill.jar        mainDexClasses
aidl                          lib             mainDexClasses.rules
arm-linux-androideabi-ld      libbcc.so       mipsel-linux-android-ld
bcc_compat                    libbcinfo.so    NOTICE.txt
dexdump                       libclang.so     renderscript
dx                            libc++.so       runtime.properties
i686-linux-android-ld         libLLVM.so      source.properties
jack.jar                      llvm-rs-cc      zipalign
$ pwd
/opt/androidsdk/build-tools/21.1.1
$
```

Figure 3-9D Listing of particular version of build tools.

Figure 3-9D shows a particular version of the Android SDK's build tools. These tools are typically used via the build process rather than directly. Building an Android project entails a particular version of the build tools. Over time, the Android project's target API level and build tool versions may undergo updates, as we will cover later in this chapter.

3.8.3 Installing Build Tools (*gradle* and *ant*)

Now that you have installed your stand-alone Android SDK, the next step is to install *Gradle* as well as *Apache Ant*, which are two popular build tools. *Apache Ant* can be downloaded from http://ant.apache.org/bindownload.cgi. In case you already have *ant* installed on your development machine, that should suffice. You will likely find that the Android classic build system that is based on *ant* generally shows no sensitivity toward the *ant* version. This book covers both the "classic" and the "new" Android project structure and build system. The new and improved Android build system is based on *gradle*.

The following are useful reference documentation for Android projects and builds:

http://developer.android.com/tools/projects/projects-cmdline.html
http://tools.android.com
http://tools.android.com/tech-docs

gradle can be downloaded from https://www.gradle.org. You may find that the Android new build system shows some sensitivity toward the version of *gradle*. Please be guided by the latest information available at the android.com links above. At the time of writing, I found *gradle* 2.1 to be a good choice.

I have several versions of *gradle* under the */opt* area. I am able to switch between various versions as needed in order to support various projects that may have particular dependency constraints. Maintaining these versions separately under discrete home directories keeps them isolated and easy enough to switch in or out, as needed. I recommend avoiding using *sudo apt-get* or *rpm*-based installs as these get installed at a system level.

3.8.4 Setting up environment variables (Java, Android SDK, *gradle* and *ant*)

Once you have *ant* and *gradle* installed, it would be ideal to set some environment variables in your *.profile* or equivalent. The following clipping indicates the various environment variable settings that are in place on my development machine due to entries in my .profile file with entries such as "*export JAVA_HOME=/opt/jdk1.7*" and more as shown in the box below.

```
ANDROID_HOME=/opt/androidsdk
GRADLE_HOME=/opt/gradle
ANT_HOME=/opt/ant
JAVA_HOME=/opt/jdk1.7
```

You will also need an entry to add $JAVA_HOME/bin, *$ANDROID_HOME/tools*, *$ANDROID_HOME/platform-tools*, *$ANDROID_HOME/build-tools/21.1.2*, *$ANT_HOME/bin*, and *$GRADLE_HOME/bin* to the beginning of your *PATH*—as shown below.

```
export PATH=$JAVA_HOME/bin:$ANDROID_HOME/tools:$ANDROID_HOME/platform-tools:$ANDROID_HOME/build-tools/21.1.2:$ANT_HOME/bin:$GRADLE_HOME/bin:$PATH
```

You will need to set these manually in your current shell by executing your .profile preceded by the "." as in ". ~/.profile".

After this, the commands *java -version*, *gradle -version*, and *ant -version* should all work and display the expected respective version information (Figure 3-10).

```
$ java -version
java version "1.7.0_45"
Java(TM) SE Runtime Environment (build 1.7.0_45-b18)
Java HotSpot(TM) 64-Bit Server VM (build 24.45-b08, mixed mode)
$ gradle -version

------------------------------------------------------------
Gradle 2.1
------------------------------------------------------------

Build time:   2014-09-08 10:40:39 UTC
Build number: none
Revision:     e6cf70745ac11fa943e19294d19a2c527a669a53

Groovy:       2.3.6
Ant:          Apache Ant(TM) version 1.9.3 compiled on December 23 2013
JVM:          1.7.0_45 (Oracle Corporation 24.45-b08)
OS:           Linux 3.11.0-14-generic amd64

$ ant -version
Apache Ant(TM) version 1.9.2 compiled on July 8 2013
$ █
```

Figure 3-10 java, *gradle*, and ant commands, versions.

3.8.5 Android (Lollipop) Development Device setup

Once you have your Android development environment setup, you will need to setup your Android Lollipop device for development and debugging—if you have not done so already. The detailed instructions are available at http://developer.android.com/tools/device.html.

You will essentially need to access your device's Android *Settings* → *System* → *About* (Phone/Tablet/device). Under *About* (Phone/Tablet/Device), you will find the item *Build number* toward the bottom. You will need to tap on *Build number* seven times. After that, you will find a new item under *Settings*, namely, *Developer options*. Under *Developer options*, you will need to check *USB debugging*. You will then need to connect your Android device to your development machine using a USB cable.

The steps for enabling debugging on an Android Wear device over USB have been covered in Section 6.7.2.3. Although the screens vary, the steps for enabling Developer options and USB debugging via *Settings* are similar. Also, Section 6.8.1 covers the steps for plugging in an Android device via the USB cable and diagnosing issues. Please feel free to refer to these sections if you run into any issues.

In case you intend to use an emulator running a Lollipop Android Virtual Device (AVD), you can create an AVD using Tools → Manage AVDs. Detailed instructions are available at https://developer.android.com/tools/devices/managing-avds.html.

Section 6.7.2.1 in this book happens to cover AVD creation for Wear devices, and it turns out that the steps for creating an AVD for an Android 5 phone or tablet simply entail selecting an appropriate Android 5 phone or tablet device in the *Device* drop down.

After that, running the *adb* devices command should list your connected devices.

This next section requires that you can list your connected Android Lollipop device using the command adb devices, as shown in Figure 3-11.

```
$ adb kill-server
$ adb devices
* daemon not running. starting it now on port 5037 *
* daemon started successfully *
List of devices attached
00ea1b9d        device

$ adb shell
shell@flo:/ $ █
```

Figure 3-11 Lollipop device adb debugging, adb devices.

3.8.5.1 Creating a new Android project (classic / ant) There is no better way to verify an Android development environment than writing our first "Hello Lollipop World" App. It takes just two steps to create an Android App and install it on your device. The android create project command provides the ability to create a new Android App/ project via the command line.

Whenever you create an Android App, you are required to **target** it to an *Android API level*. The API level represents the version of the Android OS, expressed as a numeric integer value such as 1 through 21, representing, respectively, Android 1.0 through Android 5.0. The preview for Lollipop code named "L preview" had the API level of "L"—as a temporary exception to the numeric API level scheme.

The mapping of Android versions, API levels, and code names (such as Lollipop, Jellybean, etc.) is available at http://developer.android.com/guide/topics/manifest/uses-sdk-element.html.

Figure 3-12 shows a screenshot from the table that you can find toward the bottom of the page linked above. As you will notice, there are three ways to refer to an Android platform version; for instance, Android platform **version 5.0** has the **API level** of **21** and the **code name** of **Lollipop**.

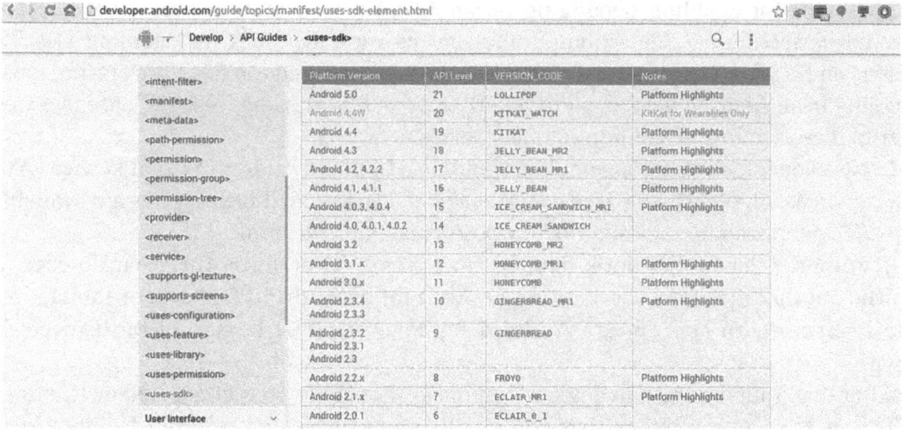

developer.android.com/guide/topics/manifest/uses-sdk-element.html

Develop > API Guides > <uses-sdk>

<intent-filter>
<manifest>
<meta-data>
<path-permission>
<permission>
<permission-group>
<permission-tree>
<provider>
<receiver>
<service>
<supports-gl-texture>
<supports-screens>
<uses-configuration>
<uses-feature>
<uses-library>
<uses-permission>
<uses-sdk>
User Interface

Platform Version	API Level	VERSION_CODE	Notes
Android 5.0	21	LOLLIPOP	Platform Highlights
Android 4.4W	20	KITKAT_WATCH	KitKat for Wearables Only
Android 4.4	19	KITKAT	Platform Highlights
Android 4.3	18	JELLY_BEAN_MR2	Platform Highlights
Android 4.2, 4.2.2	17	JELLY_BEAN_MR1	Platform Highlights
Android 4.1, 4.1.1	16	JELLY_BEAN	Platform Highlights
Android 4.0.3, 4.0.4	15	ICE_CREAM_SANDWICH_MR1	Platform Highlights
Android 4.0, 4.0.1, 4.0.2	14	ICE_CREAM_SANDWICH	
Android 3.2	13	HONEYCOMB_MR2	
Android 3.1.x	12	HONEYCOMB_MR1	Platform Highlights
Android 3.0.x	11	HONEYCOMB	Platform Highlights
Android 2.3.4 Android 2.3.3	10	GINGERBREAD_MR1	Platform Highlights
Android 2.3.2 Android 2.3.1 Android 2.3	9	GINGERBREAD	
Android 2.2.x	8	FROYO	Platform Highlights
Android 2.1.x	7	ECLAIR_MR1	Platform Highlights
Android 2.0.1	6	ECLAIR_0_1	

Figure 3-12 Android API levels, code names, and OS versions.

The command android list targets lists the API level targets that have been installed on your development machine via the Android SDK Manager.

```
Available Android targets:
----------
id: 1 or "android-14"
     Name: Android 4.0
     Type: Platform
     API level: 14
     Revision: 4
     Skins: HVGA, QVGA, WQVGA400, WQVGA432, WSVGA, WVGA800 (default), WVGA854, W
XGA720, WXGA800
 Tag/ABIs : default/armeabi-v7a
----------
id: 2 or "android-15"
     Name: Android 4.0.3
     Type: Platform
     API level: 15
     Revision: 5
     Skins: HVGA, QVGA, WQVGA400, WQVGA432, WSVGA, WVGA800 (default), WVGA854, W
XGA720, WXGA800
 Tag/ABIs : default/armeabi-v7a, default/mips, default/x86
----------
id: 3 or "android-16"
     Name: Android 4.1.2
     Type: Platform
--More--
```

Figure 3-13A adb list targets, partial/initial output.

Figure 3-13A shows the initial output of the adb list targets command. The latter portion of the same output is shown in the next command.

Figure 3-13B shows the latter portion of the output of the adb list targets command. One of the nuances of the Android SDK is that it maintains a local, relative ID for each platform API level/target that you have installed via the SDK Manager.

```
            API for Google Maps
     Skins: HVGA, QVGA, WQVGA400, WQVGA432, WSVGA, WVGA800 (default), WVGA854, W
XGA720, WXGA800, WXGA800-7in
 Tag/ABIs : default/x86
----------
id: 17 or "Google Inc.:Google APIs:21"
     Name: Google APIs
     Type: Add-On
     Vendor: Google Inc.
     Revision: 1
     Description: Android + Google APIs
     Based on Android 5.0.1 (API level 21)
     Libraries:
      * com.google.android.media.effects (effects.jar)
          Collection of video effects
      * com.android.future.usb.accessory (usb.jar)
          API for USB Accessories
      * com.google.android.maps (maps.jar)
          API for Google Maps
     Skins: HVGA, QVGA, WQVGA400, WQVGA432, WSVGA, WVGA800 (default), WVGA854, W
XGA720, WXGA800, WXGA800-7in, AndroidWearRound, AndroidWearSquare, AndroidWearRo
und, AndroidWearSquare
 Tag/ABIs : google_apis/armeabi-v7a, google_apis/x86, google_apis/x86_64
$
```

Figure 3-13B adb list targets, partial/terminal output.

In my local development environment, the output of adb list targets tells me that my local, relative ID for the API level 21 is id: 17.

```
id: 17 or "Google Inc.:Google APIs:21"
   Name: Google APIs
   Type: Add-On
   Vendor: Google Inc.
   Revision: 1
   Description: Android + Google APIs
   Based on Android 5.0.1 (API level 21)

...
```

Once I have determined the relative/local ID for my target platform of interest (Android 5/Lollipop), I am ready to proceed with creating my first Android App/project.

I created a new directory for my project and proceeded to issue the command as under:

```
~projects$ cd 0hellol
~0hellol$ android create project --path . --name 0hellol --package io.wearbook.hellol --
activity HelloLollipopActivity --target 17
```

You will notice that any Android project needs a package name (which by convention, you align with an Internet domain name that you own, such as wearbook.io), an Activity class name, and a target using the relative, local ID (Figure 3-14A).

If the above command executed successfully, you are one step away from installing your first Android App on your Android device. The command shown below will build your App and install it to your connected Android Lollipop device (Figure 3-14B).

```
~0hellol$ ant clean debug install
```

Once you have successfully built and installed your first App to your Lollipop device, you can launch your App either by locating the App on your device or simply by launching a command as under (Figure 3-14C):

```
~0hellol$adb shell am start -a android.intent.action.MAIN -n
io.wearbook.hellol/.HelloLollipopActivity
```

The *am* command helps you automate the starting of your App without having to touch the screen.

The following are screenshots of the commands above, executed from the command line.

```
~/projects/0hellol $ android create project --path . --name 0hellol --package io.wearbook.he
llol --activity HelloLollipopActivity --target 17
Created directory /home/sanjay/projects/0hellol/src/io/wearbook/hellol
Added file ./src/io/wearbook/hellol/HelloLollipopActivity.java
Created directory /home/sanjay/projects/0hellol/res
Created directory /home/sanjay/projects/0hellol/bin
Created directory /home/sanjay/projects/0hellol/libs
Created directory /home/sanjay/projects/0hellol/res/values
Added file ./res/values/strings.xml
Created directory /home/sanjay/projects/0hellol/res/layout
Added file ./res/layout/main.xml
Created directory /home/sanjay/projects/0hellol/res/drawable-xhdpi
Created directory /home/sanjay/projects/0hellol/res/drawable-hdpi
Created directory /home/sanjay/projects/0hellol/res/drawable-mdpi
Created directory /home/sanjay/projects/0hellol/res/drawable-ldpi
Added file ./AndroidManifest.xml
Added file ./build.xml
Added file ./proguard-project.txt
~/projects/0hellol $ ant clean debug install
```

Figure 3-14A android create project, ant install.

Figure 3-14D shows a screenshot of the "Hello" App running on a Lollipop device. You may notice that it does have a good resolution and has an "outdated" look. It turns out that although we have targeted the App to the "current" Android 5 platform, the *AndroidManifest.xml* file located in the project's home directory reveals upon examination that it does not have a "minimum SDK level" set—therefore, it defaults to "1," which is the very first version of Android 1.0. The API and features that an Android App has access to are determined by the "minimum" and "target" SDK levels.

```
 [zipalign] Running zip align on final apk...
     [echo] Debug Package: /home/sanjay/projects/0hellol/bin/0hellol-debug.apk
[propertyfile] Creating new property file: /home/sanjay/projects/0hellol/bin/build.prop
[propertyfile] Updating property file: /home/sanjay/projects/0hellol/bin/build.prop
[propertyfile] Updating property file: /home/sanjay/projects/0hellol/bin/build.prop
[propertyfile] Updating property file: /home/sanjay/projects/0hellol/bin/build.prop

-post-build:

debug:

install:
     [echo] Installing /home/sanjay/projects/0hellol/bin/0hellol-debug.apk onto default emul
ator or device...
     [exec] 891 KB/s (37782 bytes in 0.041s)
     [exec]      pkg: /data/local/tmp/0hellol-debug.apk
     [exec] Success

BUILD SUCCESSFUL
Total time: 6 seconds
~/projects/0hellol $
```

Figure 3-14B ant install successful.

```
$ adb shell am start -a android.intent.action.MAIN -n -i io.wearbook.hellol/.Hel
loLollipopActivity
```

Figure 3-14C adb shell am.

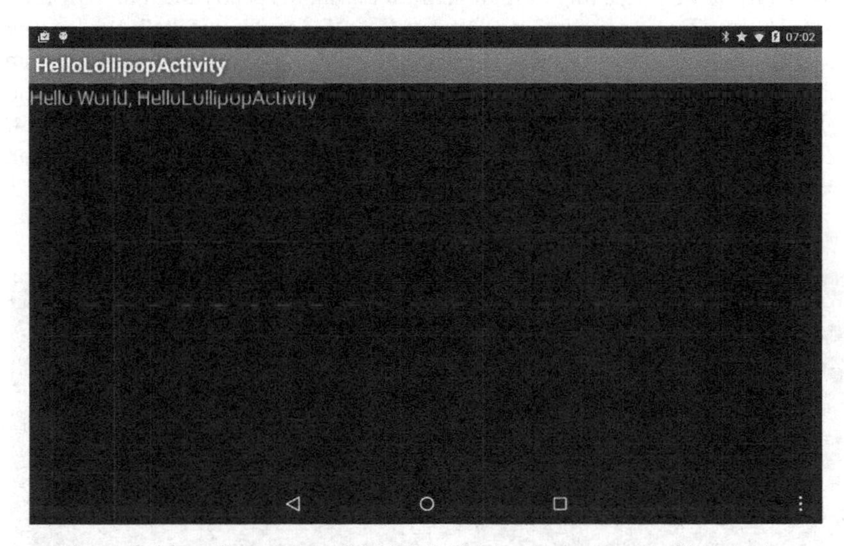

Figure 3-14D Screenshot of the "Hello" App running on a Lollipop tablet.

By editing the *AndroidManifest.xml* and adding the line within the *manifest* element as under:

```
<uses-sdk android:minSdkVersion="21" />
```

and repeating the compiling, installing, and running steps will show you a much improved, high-resolution look. By setting the minimum SDK level to the Android 5 platform, your App cannot be installed on a prior version of Android such as 4.4/KitKat.

```
AndroidManifest.xml + (~/projects/0hellol) - VIM
File Edit View Search Terminal Help
<?xml version="1.0" encoding="utf-8"?>
<manifest xmlns:android="http://schemas.android.com/apk/res/android"
      package="io.wearbook.hellol"
      android:versionCode="1"
      android:versionName="1.0">
    <uses-sdk android:minSdkVersion="21" />
    <application android:label="@string/app_name" android:icon="@drawable/ic_launcher">
        <activity android:name="HelloLollipopActivity"
                  android:label="@string/app_name">
            <intent-filter>
                <action android:name="android.intent.action.MAIN" />
                <category android:name="android.intent.category.LAUNCHER" />
            </intent-filter>
        </activity>
    </application>
</manifest>
~
~
~
~
                                                              6,43          All
```

Figure 3-14E Adding a min SDK level.

Figure 3-14E shows the editing of *AndroidManifest.xml* to specify the min SDK level to 21 (Lollipop). By specifying a minimum SDK level of 21, for this "Hello" App, it cannot be installed or run on Android 4.4 (KitKat). In practice, you will need to choose a minimum SDK level coincident with the most "ancient" Android platform that you intend your App to support. Practically, that may correspond to, say, Android 2.3.3 or Android 4.0.3 or Android 4.4. Most likely, it will not be the very first version of Android released in 2008—which is the default value that kicks in, in the absence of an explicit value set in your *AndroidManifest.xml* (Figure 3-14F).

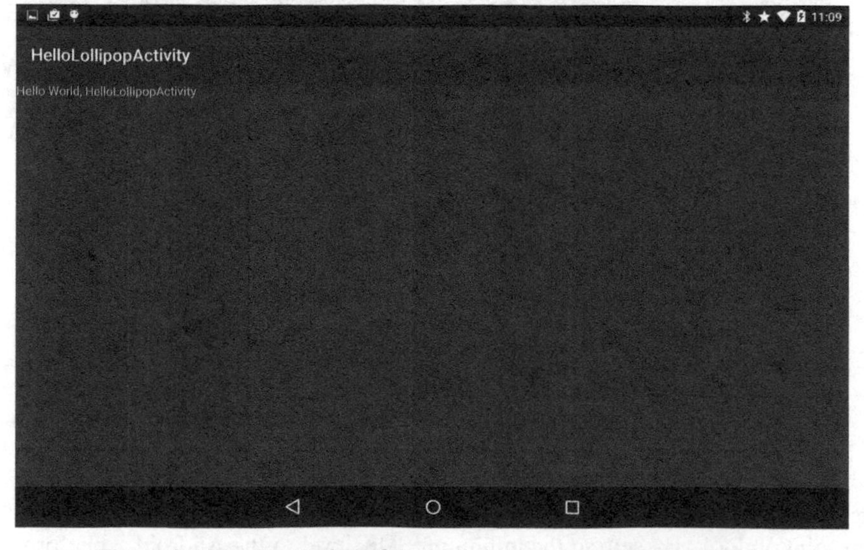

Figure 3-14F Screenshot of the Hello* App with an improved look.

You may explore the contents of this newly created project tree; it represents a "classic" Android project tree.

3.8.5.2 Creating a new Android project (new / *gradle*) In this section, we will create a new Android project using the "new" build system that is based on *gradle*. Once again, we will use the *android create project* command with a couple of flags to opt for a *gradle*-based project. This example has been tested to work using *gradle* 1.12. You may find version sensitivities using a different version of *gradle* or the plug-in version.

The following steps will create a new project, build, install, and run it on your device:

```
android create project --path . --name 0hellol --package io.wearbook.hellolg --activity
HelloLollipopGradleActivity --target 17 --gradle  --gradle-version 0.11.1

gradle installDebug

adb shell am start -a android.intent.action.MAIN -n
io.wearbook.hellolg/.HelloLollipopGradleActivity
```

Figure 3-15A shows the android create command in action—with the *gradle* options and flag enabled. The *gradle installDebug* command builds and installs the App (Figures 3-15B and 3-15C).

```
~/projects/1hellol $ android create project --path . --name 0hellol --package io.wearbook.hellol --a
ctivity HelloLollipopActivity --target 17 --gradle   --gradle-version 0.11.1
Created directory /home/sanjay/projects/1hellol/src/main/java
Created directory /home/sanjay/projects/1hellol/src/main/java/io/wearbook/hellol
Added file ./src/main/java/io/wearbook/hellol/HelloLollipopActivity.java
Created directory /home/sanjay/projects/1hellol/src/androidTest/java
Created directory /home/sanjay/projects/1hellol/src/androidTest/java/io/wearbook/hellol
Added file ./src/androidTest/java/io/wearbook/hellol/HelloLollipopActivityTest.java
Created directory /home/sanjay/projects/1hellol/src/main/res
Created directory /home/sanjay/projects/1hellol/src/main/res/values
Added file ./src/main/res/values/strings.xml
Created directory /home/sanjay/projects/1hellol/src/main/res/layout
Added file ./src/main/res/layout/main.xml
Created directory /home/sanjay/projects/1hellol/src/main/res/drawable-xhdpi
Created directory /home/sanjay/projects/1hellol/src/main/res/drawable-hdpi
Created directory /home/sanjay/projects/1hellol/src/main/res/drawable-mdpi
Created directory /home/sanjay/projects/1hellol/src/main/res/drawable-ldpi
Added file ./src/main/AndroidManifest.xml
Added file ./build.gradle
Created directory /home/sanjay/projects/1hellol/gradle/wrapper
~/projects/1hellol $ gradle installDebug
```

Figure 3-15A android create *gradle*-based project.

```
:compileDebugNdk
:preDebugBuild
:checkDebugManifest
:prepareDebugDependencies
:compileDebugAidl
:compileDebugRenderscript
:generateDebugBuildConfig
:generateDebugAssets UP-TO-DATE
:mergeDebugAssets
:generateDebugResValues
:generateDebugResources
:mergeDebugResources
:processDebugManifest
:processDebugResources
:generateDebugSources
:compileDebugJava
:preDexDebug
:dexDebug
:processDebugJavaRes UP-TO-DATE
:validateDebugSigning
:packageDebug
:zipalignDebug
:installDebug
903 KB/s (38033 bytes in 0.041s)
        pkg: /data/local/tmp/1hellol-debug.apk
Success

BUILD SUCCESSFUL

Total time: 11.849 secs
~/projects/1hellol $ adb shell am start -a android.intent.action.MAIN -n io.wearbook.hellolg/.HelloL
ollipopGradleActivity
```

Figure 3-15B adb shell am command, to launch activity.

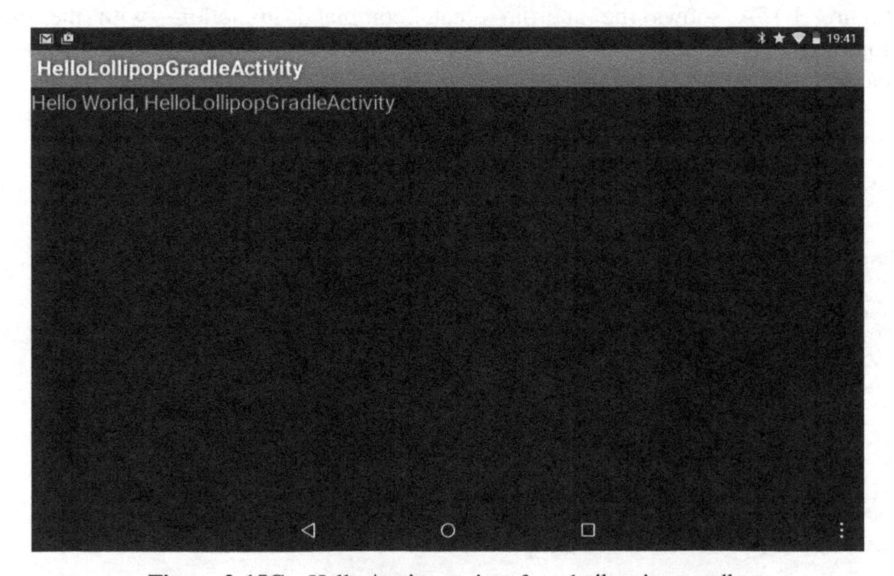

Figure 3-15C Hello App's user interface, built using *gradle*.

3.8.6 Installing *Android Studio* "IDE"

The *Eclipse*-based Android development IDE has been replaced by the new *Android Studio IDE*, going forward. This book does not cover *Eclipse* at all; however, the stand-alone *Android SDK* installation that I covered in the earlier section is an approach that is aligned with using the command line and also using any IDE, including but not limited to *Android Studio*. The command line-based build is useful and eventually needed for build automation, continuous integration, configuration management, and testing. Deeper awareness and understanding of the core *Android SDK* tools can make it easier to switch to a different IDE, with lesser effort. This book is somewhat IDE agnostic; however, *Android Studio* is the recommended and predominantly used IDE in this book. *Netbeans* (www.netbeans.org) is a fine Java IDE and has had a plug-in for Android since some time; recently, support has been added for *gradle*-based Android projects. There is no indication that Android *gradle* support is coming to Eclipse anytime soon, but such matters are not easy to predict.

The steps that I took to carry out the installation of *Android Studio* on my development machine were straightforward. I downloaded the Linux version of the *Android Studio* binary zip archive from http://developer.android.com/sdk/index.html#Other (Figure 3-16A).

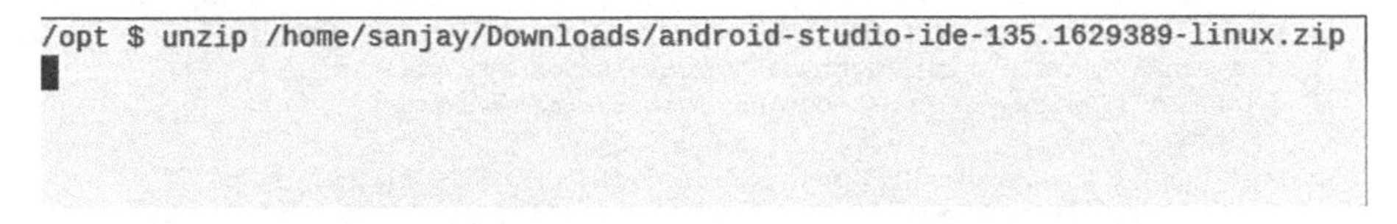

/opt $ unzip /home/sanjay/Downloads/android-studio-ide-135.1629389-linux.zip

Figure 3-16A Unzip Android Studio download.

I unzipped the archive into */opt* and then kicked started *Android Studio* via the command below (Figure 3-16B):

```
/opt/android-studio/bin/studio.sh &
```

(I also noticed that an *Android Studio* shortcut had been created on my desktop.)

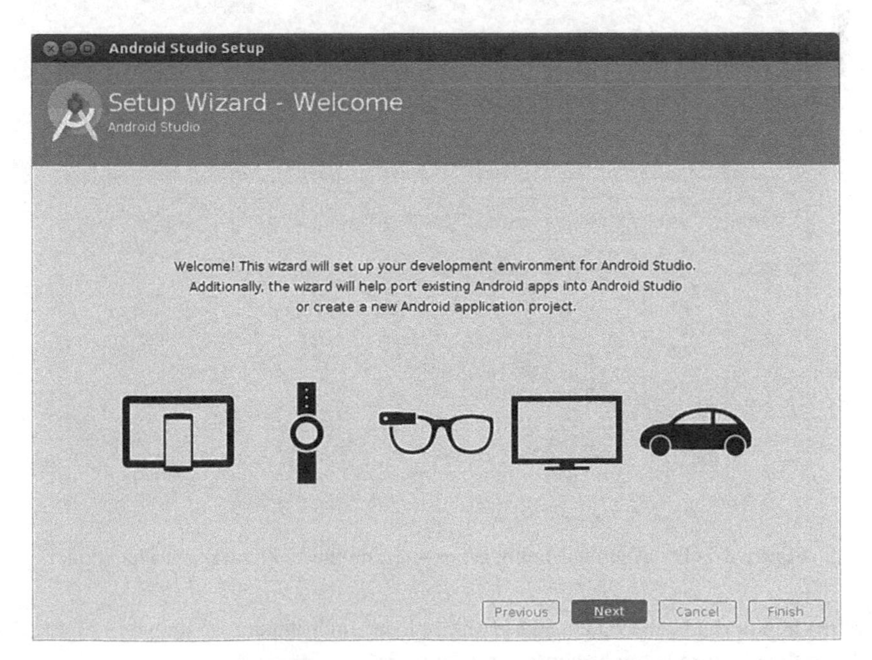

Figure 3-16B Android Studio, welcome screen.

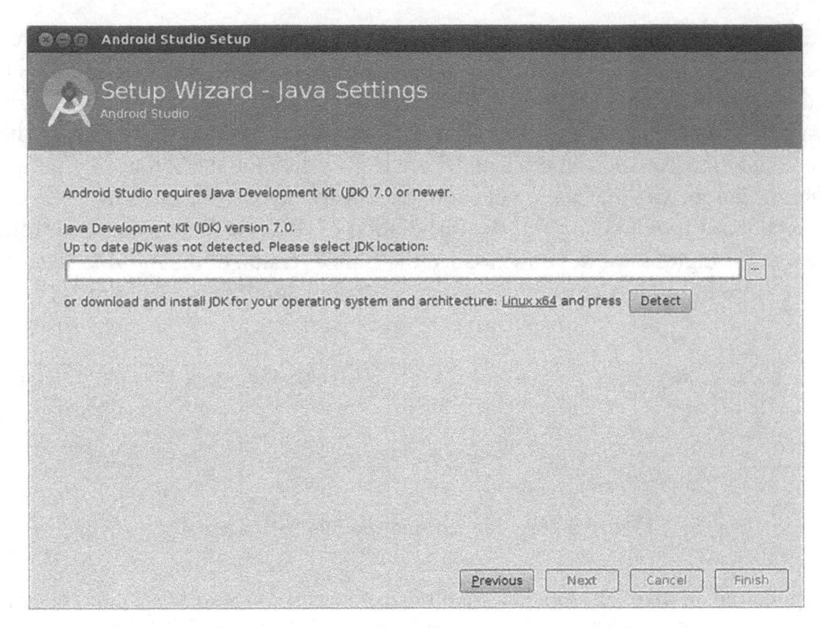

Figure 3-16C Android Studio, setup—JDK.

I started *Android Studio* and was prompted for the *Java 7* installation location, as shown in Figure 3-16C.

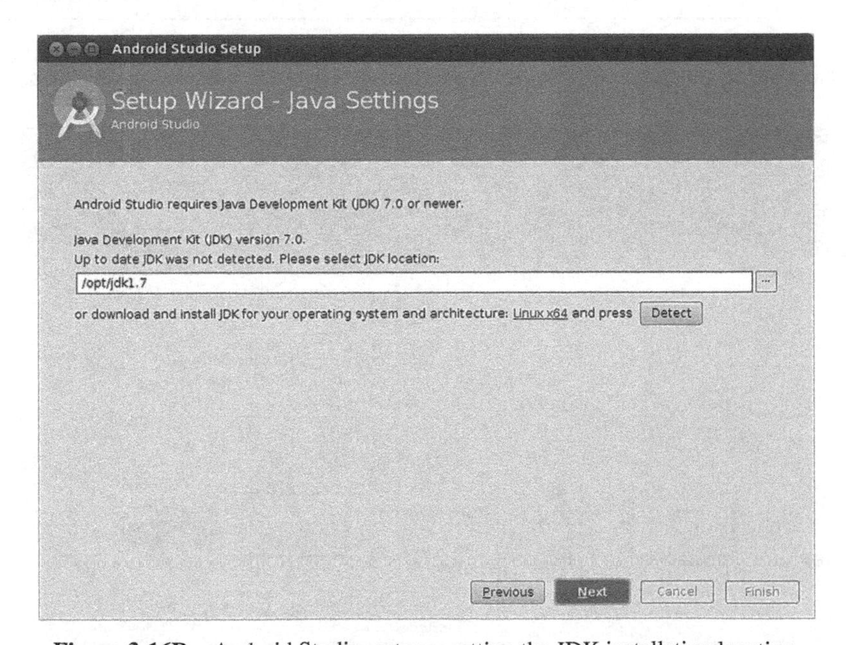

Figure 3-16D Android Studio, setup—setting the JDK installation location.

I set the appropriate *Java JDK* location on my local environment, as shown in Figure 3-16D. I chose the Custom setup as shown in Figure 3-16E

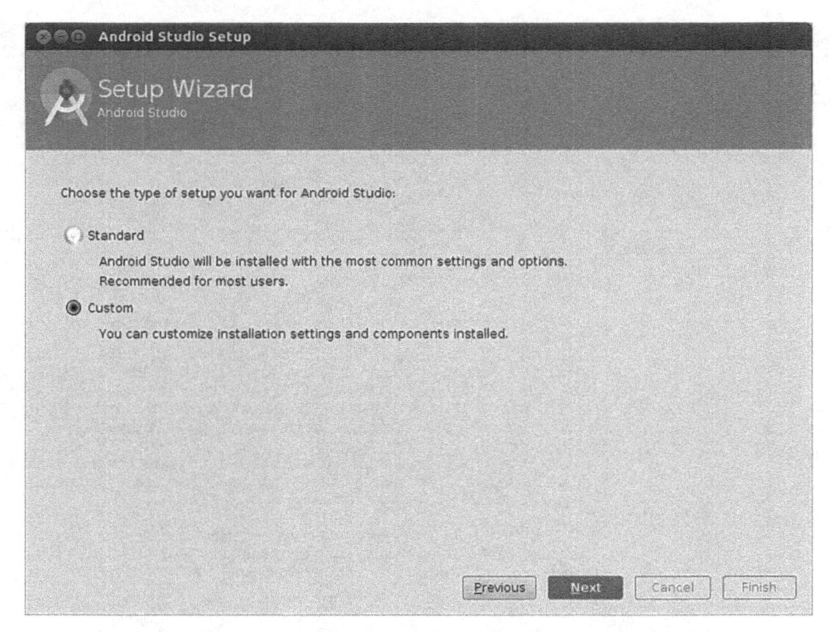

Figure 3-16E Android Studio, setup—standard or custom.

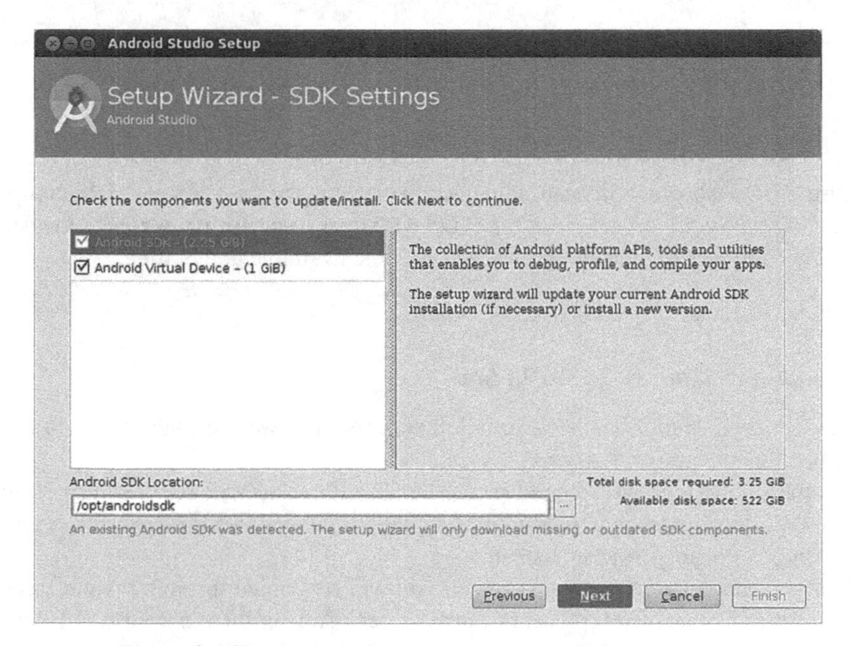

Figure 3-16F Android Studio, setup—Android SDK location.

Figure 3-16F shows the next screen in the flow, which detects/or lets you set the intended Android SDK location. It does not download the Android SDK separately; however, it will run the Android SDK Manager to check if any updates are needed.

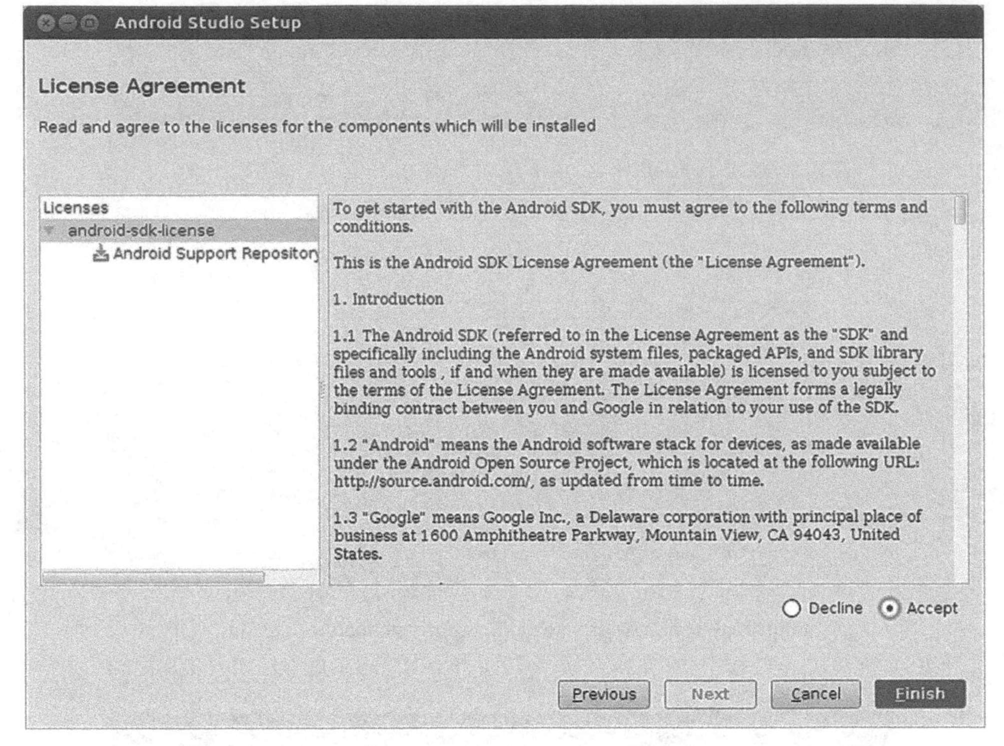

Figure 3-16G Android Studio, setup—Android SDK update, license agreement.

Figure 3-16G shows a license agreement pertinent to an Android SDK component update. *Android Studio* uses the *Android SDK Manager* to perform updates to the Android SDK. New Android component updates can become available at any time.

3.8.7 *Android Studio*: Hello World App

Now that *Android Studio* has been installed successfully, we are ready to write our first Android Lollipop App using *Android Studio*.

After the setup steps we covered in the last section, you will see a screen similar to Figure 3-17A. It provides you with options for creating a new Android project, opening an existing one, importing projects, and so on.

When you create a new Android project, you will be guided through several screens to specify the Activity (which represents a new screen) and set the Android API level/target for your App.

Figure 3-17B shows the screen where you provide the name, layout, and such information about the main Activity for your App. This is the screen that is displayed when your App starts.

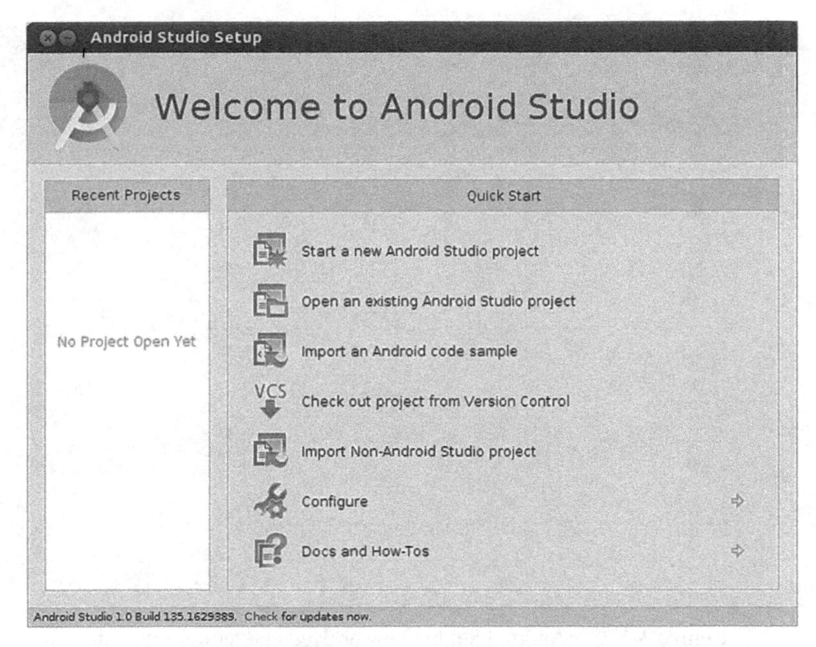

Figure 3-17A Android Studio, quick start.

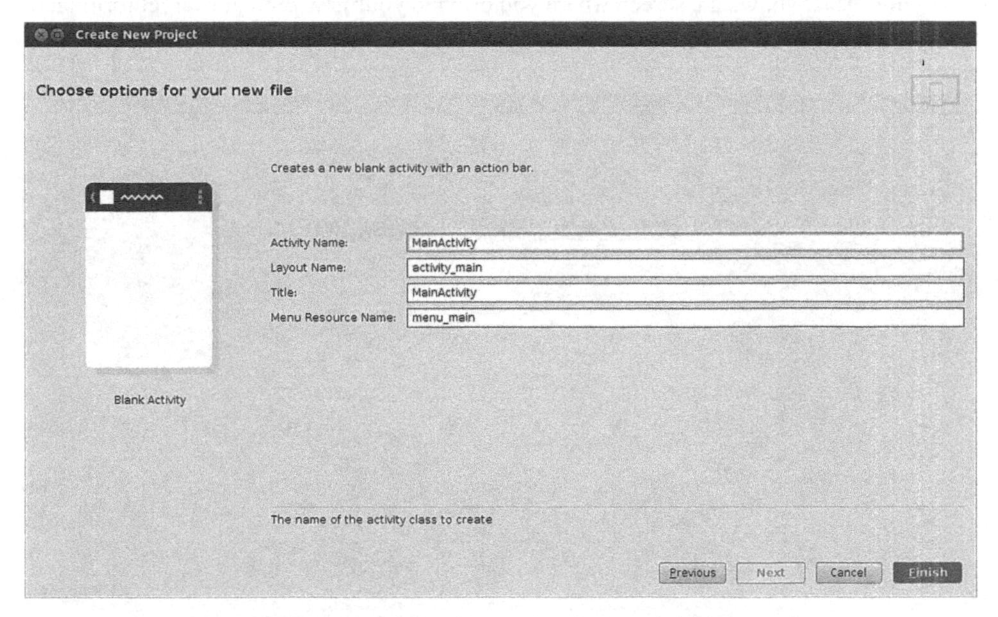

Figure 3-17B Android Studio, new project's activity information.

Create New Project

New Project
Android Studio

Select the form factors your app will run on

Different platforms require separate SDKs

Phone and Tablet
Minimum SDK API 21: Android 5.0 (Lollipop)
Lower API levels target more devices, but have fewer features available. By targeting API 21 and later, your app will run on approximately 0.0% of the devices that are active on the Google Play Store. Help me choose.
TV
Minimum SDK
Wear
Minimum SDK
Glass
Minimum SDK

Previous Next Cancel Finish

Figure 3-17C Android Studio, new project's target information.

Figure 3-17C shows the screen where you provide your new project's target form factor and target API information.

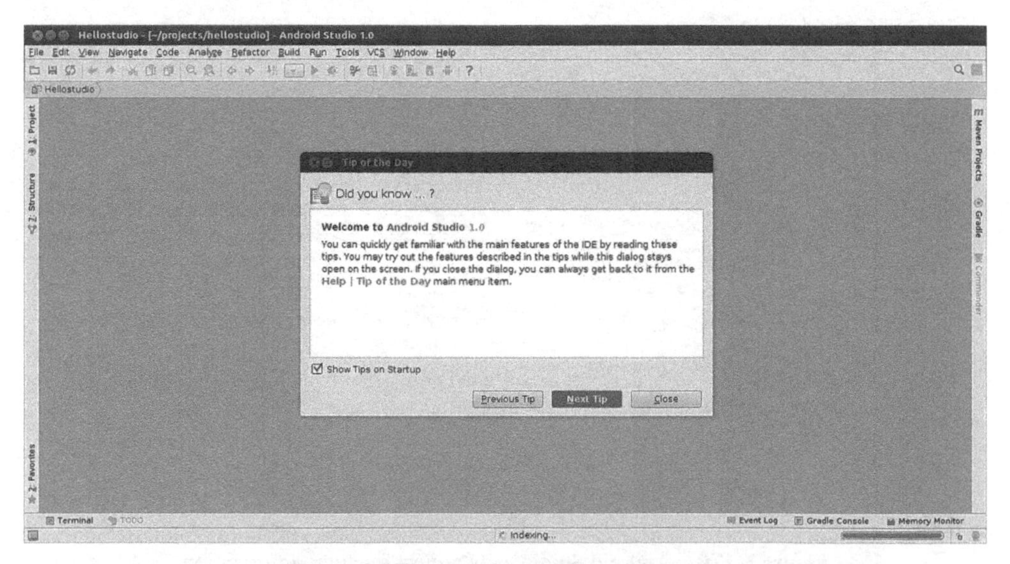

Figure 3-17D Android Studio, new project.

Figure 3-17D shows the newly created project. Toward the top left, you will find that you can access the project's tree in different modes and browse through the project's code and resources.

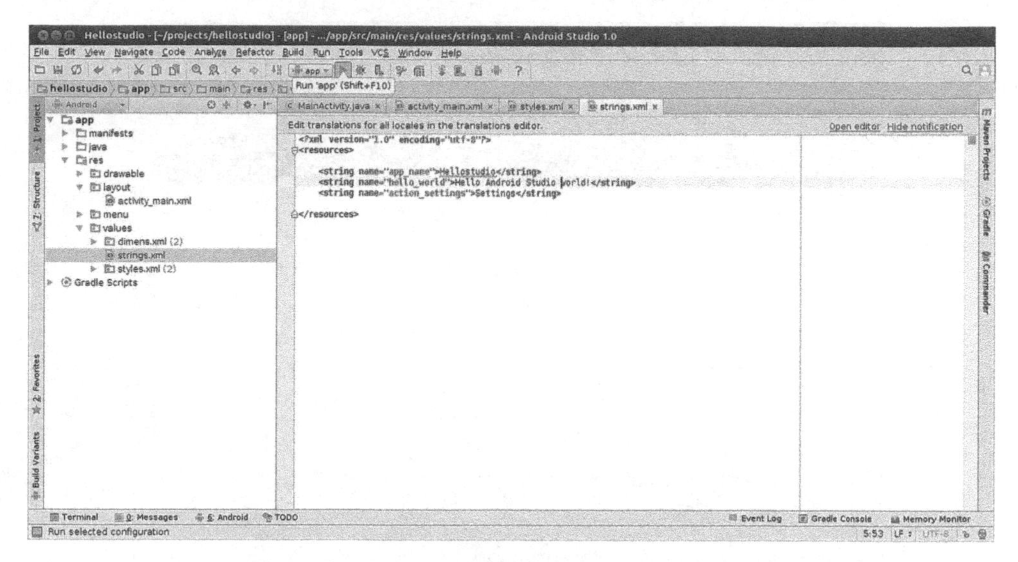

Figure 3-17E Android Studio, editing and building.

Figure 3-17E shows project's code and resources that can be edited. There is a green Play/Run App button on the top bar. Clicking on this Run App button will build and run the App.

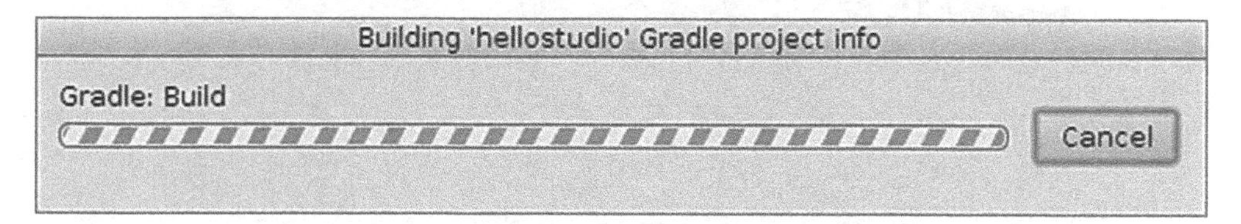

Figure 3-17F Android Studio, building and running.

Figure 3-17F shows the dialog that pops up upon clicking on the Run App button. After the project compiles successfully, you will be prompted to choose a connected device to install and run the App on.

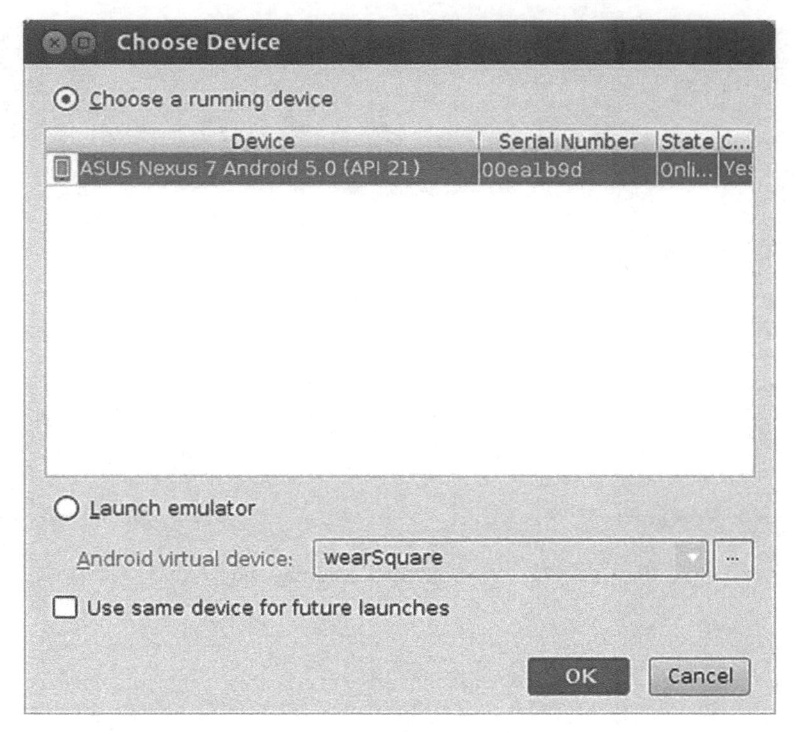

Figure 3-17G Android Studio, choosing a connected device.

Figure 3-17G shows the *Choose Device* dialog, which prompts you to choose the device on which you would like to install and run your App on.

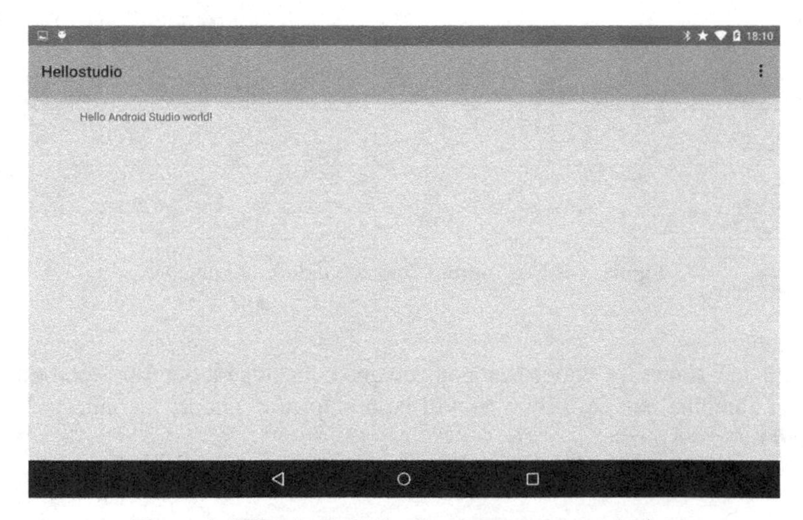

Figure 3-17H Android Studio, App running on device.

Figure 3-17H shows the Hello Studio App running on a Nexus 7 device.

3.8.8 Configuring *Android Studio*

Android Studio can be customized per your needs—you will find a *Configure* option at the bottom of the *Quick Start* area in the main/initial screen.

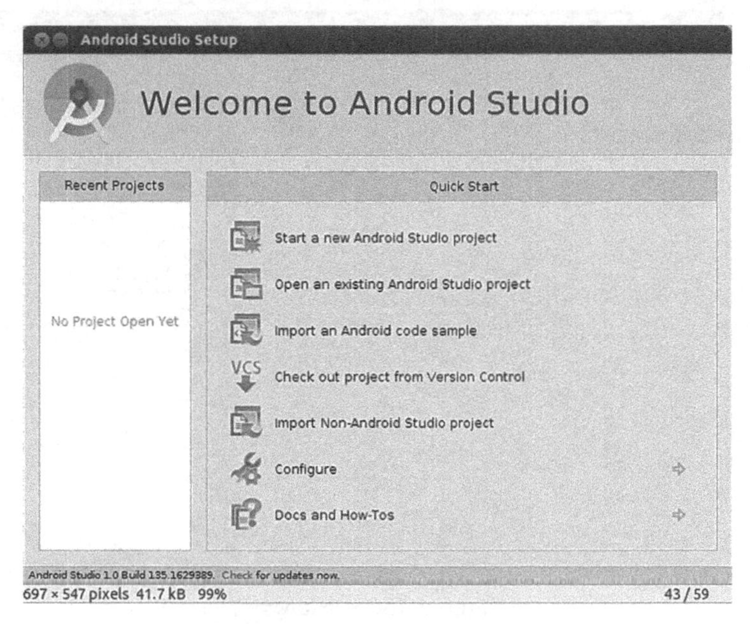

Figure 3-18A Android Studio, configure.

Figure 3-18A shows the *Configure* option toward the bottom of the *Quick Start* area.

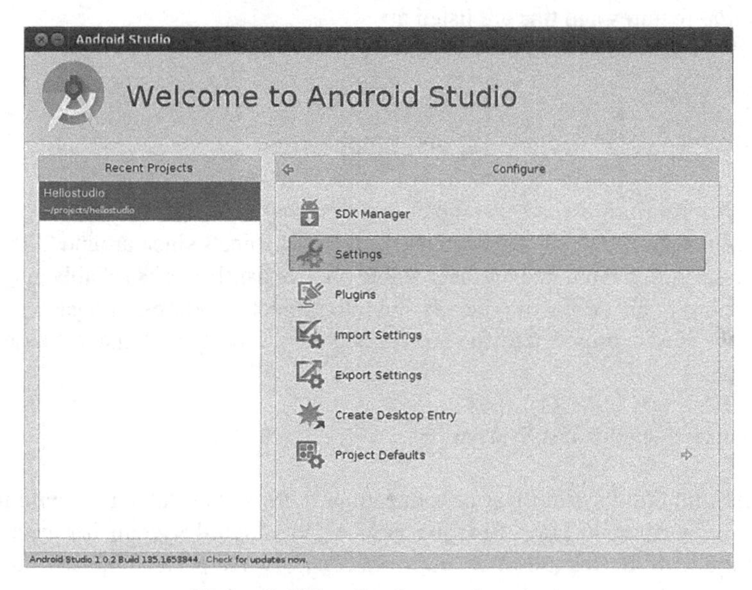

Figure 3-18B Configure sub options.

Figure 3-18B shows the various options under *Configure*, such as SDK Manager (which launches the Android SDK Manager that we are already familiar with) and Settings (which helps you customize and personalize the code styles, compiler settings, etc.).

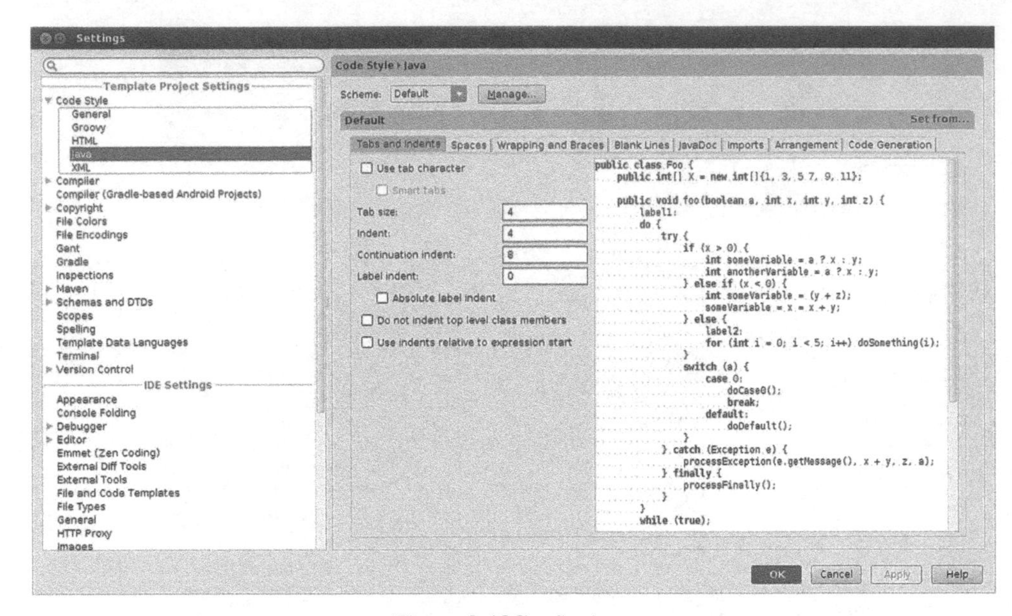

Figure 3-18C Settings.

Figure 3-18C shows the Settings area for Java code style.

As you get familiar with *Android Studio*, you will notice several useful features in action such as its top-notch UI designer, language/internationalization features and build variants. *Android Studio* features and tips are listed at:

https://developer.android.com/sdk/installing/studio-tips.html

3.9 Android "Classic" project tree and build system

The classic Android project tree is aligned with ant-based builds. Ant is a cross-platform build tool from Apache, which has been used for Java projects since around 2000. You will find that the ant-based build is extremely stable and robust. It works reliably across a wide range of versions of ant. You will typically run into projects, libraries, or legacy projects that already use the classic project tree and are aligned more closely with the *ant*-based build.

3.10 Android "New" Build System

The new Android build system was announced and released as "alpha" around mid-2013. At the time of writing in late 2014, the new Android build system has been formally released along with the release of the *Android Studio* version 1.0. The new build system addresses many aspects and complexities off the shelf. There are several new concepts that have been introduced in the new build system including build variants, product flavors, and build types, which are supported exclusively—at the time of writing—by the new *gradle* build system and *Android Studio*. Going forward, you will certainly find yourself using the new build system and most likely see the advantages of *Android Studio*. As it turns out, *gradle* is compatible and provides excellent integration with *ant*.

3.11 Managing Java Installations

As developers, we often need to work on multiple projects that may require differing versions of the Java SDK. If you have a solution for this situation that already works well for you, it's best to stick to that. In this section, I share some ideas on managing the Java part of your Android SDK environment. Historically, the Android SDK itself has progressively required versions of the Java SDK such as 1.5, 1.6, and 1.7.

3.11.1 Avoid sudo apt-get / rpm style installation

I find it useful to use a zip/tar base binary or an installer that creates a discrete installation directory, whenever possible. Using the sudo apt-get or rpm-based installation is really convenient for installing general tools and utilities that need to be installed system-wide. Although you may use *sudo apt-get* to install various versions of java, that form of installation potentially "spreads" the binary contents on a system-wide basis and is not an ideal approach for managing your Java installation. Besides, newer versions of Java JDK are available exclusively from the Oracle Java JDK download page. It is best to get your Java SDK from the source. Although there do exist mechanisms to manage multiple JDK versions, it's not as foolproof as using a discrete JAVA_HOME that points to a location that contains the entire Java SDK directory tree.

3.11.2 Maintain discrete Java JDK versions

You will find yourself needing different versions of Java for your various projects. I find that it works best to use discrete Java SDK install locations and name them simply as *jdk1.6*, *jdk1.7*, *jdk1.8*, and so on (Figure 3-19). Using simple names helps ensure that you can quickly validate correctness of your PATH setting with just one glance (in contrast to using the default names that go jdk1.7.0_45_* or something like that). If you need to find out the exact minor version, the command *java -version* will provide you with that information.

```
$ ls -l /opt | grep jdk
drwxr-xr-x  8 sanjay root          4096 Mar 26  2013 jdk1.6
drwxr-xr-x  8 sanjay    143        4096 Oct  8  2013 jdk1.7
drwxr-xr-x  8 sanjay    143        4096 Dec 19  2013 jdk1.8
$ █
```

Figure 3-19 Multiple Java JDK versions.

Using a discrete location for JAVA_HOME is the simplest and most effective way of isolating the various versions of JAVA SDK installed in your environment. There are certainly other effective approaches to managing your Java environment, such as using symbolic links that use simpler names for the JAVA_HOME.

3.11.3 Set JAVA_HOME in your *.profile*

I find it useful to set JAVA_HOME in my *.profile* to the path of JAVA SDK installation that I intend to use predominantly such as *jdk1.7* or otherwise.

3.11.4 Project-wise JAVA_HOME

I find it useful to create a setenv.sh file with JAVA_HOME and PATH settings according to the JAVA SDK version dependency for your particular project. This can help quickly override the setting of JAVA_HOME in your profile. Your build scripts will need to reference the desired source level of your code such as 1.6, 1.7, 1.8, and so on, which works hand in hand with the desired tools needed for compilation.

3.11.5 IDE independent build

An IDE independent, pristine build is useful for continuous integration and configuration management. Such a build is typically based on build tools such as *gradle*, *ant*, *maven*, and so on. It is a more formal build that improves the reliability and consistency of your build process and insulates from human errors that can arise if the build is dependent on human interaction with the IDE's dialogs and screens during the formal build process.

3.12 Managing Android SDK installation and updates

In the case of server-side Java applications that have been deployed into production, you can lock down on your Java version and plan your Java version upgrades with advance notice. Upgrading the Java version is typically not a top priority from a business perspective. The landscape is quite different when it comes to Android applications, assuming you have a consumer facing Android App that is being distributed via the Google Play Store. New devices with the lastest Android OS will arrive on the market place all the time, and in order to stay competitive, App developers will need to embrace the latest version of the OS and the associated API features and improvements.

3.12.1 Update your Android SDK often

In the case of Android applications, new devices and OS updates on consumer devices do get released often and on a timeline that is beyond the control of you, the App developer. Therefore, it behooves you to update your Android SDK often, say, on a weekly basis. You will find new updates, samples and the latest support libraries, and more. It will also validate that your App continues to compile and work with the latest tools. In case anything should need refactoring, you will find that the effort will tend to be marginal.

3.12.2 Target your App to the latest SDK / API level

Always target your App to the latest API level and test it out on the emulator or physical device if you have one. This will future proof your App and even get it to run faster and more efficiently. You will also be able to leverage the latest features and deliver the best user experience for your users and customers.

3.12.3 Be sure to specify a minimum SDK / API level for your App

As we saw earlier in this chapter, the minimum SDK level defaults to the very first version of Android API level 1, which was released in 2008. It limits the features and APIs that your App can avail of. It is important to strategically choose the minimum SDK/API level -as in the oldest Android OS version - that you would like to support. Periodically you may find it meaningful to review and "slide" this level upward/forward to a more current OS.

3.13 Code Samples: Android Lollipop

The source code for this book (available online as specified in the *About this Book* section under *Website*) is organized into three top-level directories:

lollipop
wear
fit

The source code under *lollipop* covers the base Android platform and has the following project subdirectories:

0hellol 1hellol 2hellostudio

Each project has a README file with relevant build instructions.

References and Further Reading

http://www.openhandsetalliance.com/
http://en.wikipedia.org/wiki/Linux
http://en.wikipedia.org/wiki/IOS
http://www.unix.org/
http://en.wikipedia.org/wiki/Linus_Torvalds
http://en.wikipedia.org/wiki/9793_Torvalds
http://content.time.com/time/specials/packages/article/0,28804,1970858_1970909_1971691,00.html
http://en.wikipedia.org/wiki/GNU_Project

http://en.wikipedia.org/wiki/Berkeley_Software_Distribution
http://en.wikipedia.org/wiki/James_Gosling
http://en.wikipedia.org/wiki/Java_(programming_language)
http://en.wikipedia.org/wiki/Android_(operating_system)
http://www.yaffs.net/
http://en.wikipedia.org/wiki/YAFFS
http://en.wikipedia.org/wiki/OpenBinder
http://elinux.org/Android_Kernel_Features
http://en.wikipedia.org/wiki/APK_(file_format)
http://en.wikipedia.org/wiki/Java_Platform,_Micro_Edition
http://en.wikipedia.org/wiki/Apache_Harmony
http://en.wikipedia.org/wiki/OpenJDK
http://en.wikipedia.org/wiki/OpenBinder
http://www.apache.org/
http://en.wikipedia.org/wiki/GNU_General_Public_License
http://en.wikipedia.org/wiki/GPL_linking_exception
http://en.wikipedia.org/wiki/Free_Java_implementations
http://en.wikipedia.org/wiki/Free_and_open-source_software
http://en.wikipedia.org/wiki/Comparison_of_open_source_and_closed_source
http://en.wikipedia.org/wiki/GNU_General_Public_License#Version_2
http://en.wikipedia.org/wiki/Interface_description_language
https://android.googlesource.com/

Chapter 4 Android SDK

4.1 Software Components, in general

There will be times when you hear the term "software component" applied to just any application class or object. Loosely speaking, the terms component, class, and object may be used interchangeably. However, there is a more specific, narrow definition and meaning of the term *component* that goes over and beyond a class or an object in the object-oriented paradigm.

Software *components* are typically characterized by strong interfaces, strong modularity, and strong separation of responsibilities between the application logic and the runtime "container" environment that houses components. The "container" is an environment that manages the full life cycle of the *components* that execute within it. The container is responsible for loading the *component*'s class, creating and destroying instances, pooling instances, injecting application and security contexts, enforcing security, providing and controlling access to resources and services, and so on. The motivation is to be able to create reusable building blocks and integrate them into a larger environment or system and achieve the goal of interoperability and seamless integration.

Whenever software is developed within a *component*-based development model, there is a greater emphasis on understanding the contractual obligations of the application developer versus the responsibilities of the runtime container provider.

Well-known component models include *JavaBeans®*, *Enterprise JavaBeans®*, *Java Remote Method Invocation* (RMI), *Servlets®*, Corba®, and so on. *JavaBeans* is one of the simplest and widely used components. *JavaBeans* strictly adhere to the well-defined contracts with respect to setters, getters, and constructors.

Wearable Android™: Android Wear & Google Fit App Development, First Edition. Sanjay M. Mishra.
© 2015 John Wiley & Sons, Inc. Published 2015 by John Wiley & Sons, Inc.

4.2 Android Application Development Model

Android application development involves a *Java SDK*-based development environment. Android developers use the *Java* programming language and the standard Sun Java JDK from Oracle in order to write Android application source code. However, the runtime ends up executing *Dalvik* Executable (DEX) code rather than *Java* bytecode.

There are many intermediate steps involved in compiling, building, and packaging an Android application. Firstly, the source code is compiled into Java class files via the Java JDK tools; after that, the *Android SDK* tools compile the Java class files into DEX files and package them into an *Android application package* (*APK*).

4.2.1 DEX file format

DEX is a bytecode format that is used to store executable code, and such bytecode is capable of being executed on an Android runtime virtual machine (VM). The DEX format is compact and was designed keeping resource constrained, small mobile devices in mind.

Whenever you build any Android project and browse through the intermediate files within the project's directory tree, you will typically find a *classes.dex* file among the intermediate files. The *classes.dex* contains the application's classes in DEX format.

```
$ find . | grep classes.dex
./app/build/intermediates/dex/debug/classes.dex
$ which dexdump
/opt/androidsdk/build-tools/21.1.2/dexdump
$ dexdump -c  app/build/intermediates/dex/debug/classes.dex
Processing 'app/build/intermediates/dex/debug/classes.dex'...
Checksum verified
$ 
```

Figure 4-1A classes.dex and dexdump.

Figure 4-1A shows such a *classes.dex* file that is typically found within any Android project's home directory, after you have successfully built the project. The figure also shows the *dexdump* command, which is one of the *Android SDK*'s build tools and can be used to verify the checksum of a *classes.dex* file. You do not need to directly execute such build tools routinely, because your *gradle*, *ant*, or IDE-based build process takes care of this for you.

4.2.2 APK file

The APK file is the artifact that is created or output from building your Android project, and it represents your application's binary. It has the extension of *.apk* and is meant to be installed on your Android device. The *.apk* file is a zip-compatible archive that contains your application's *classes.dex* file, *AndroidManifest.xml* file and resources, and binary content.

The Android application package (.apk) file has the Internet media-type identifier of:

application/vnd.android.package-archive

Each APK has exactly one package name associated with it. When you release your Android application to the *Google Play Store*, the package name serves as the ID for the application, and it is publicly visible as in the example below:

https://play.google.com/store/apps/details?id=com.pertino.connect

One of the latter steps in the process involves the signing of the Android APK using a keystore. While building your application in debug mode, the output APK typically has debug flags enabled, and it is signed by a debug keystore. While building applications in release (i.e., production) mode, the debug flags need to be turned off, and the APK will need to be signed using a release keystore. It is perfectly fine to use a self-signed certificate rather than a certificate from a certificate authority for the keystore.

When you release an application to the *Google Play Store*, all subsequent updates will need to be made by using the same keystore that was used originally. In case you lose the original keystore and password, you will not be able to release any updates to your application on the Google Play Store. It is therefore important to maintain your keystore carefully for future use.

It is recommended that released APKs be obfuscated. Obfuscation makes the code unfriendly toward reverse engineering while also making the code more compact. More information on obfuscation and Android application signing can be found at:

http://developer.android.com/tools/publishing/app-signing.html
http://developer.android.com/tools/help/proguard.html

Also, there are particular requirements when publishing an application in the *Google Play Store*, details of which can be found at:

http://developer.android.com/tools/publishing/preparing.html
http://developer.android.com/distribute/tools/launch-checklist.html
http://developer.android.com/reference/java/security/KeyStore.html

Figure 4-1B shows an example of the Android APK file that you will find within any Android project's home directory, after you have successfully built the project. The figure shows two versions of the *APK* file, one that includes the term "unaligned" in its name. *zipalign* is an archive alignment and optimization tool that reduces the memory (RAM) consumed by the running application. *zipalign* is used in one of the last steps in the build process, and it is performed after application signing. zipalign is a build tool that is part of the *Android SDK*; it is typically used via the build process, rather than directly.

```
$ find -name *.apk
./app/build/outputs/apk/app-debug.apk
./app/build/outputs/apk/app-debug-unaligned.apk
$ which zipalign
/opt/androidsdk/build-tools/21.1.2/zipalign
$ █
```

Figure 4-1B Android apk file.

Besides the APK archive, there is the AAR archive, which is the output from building a library project or module that other Android application projects can depend on.

4.2.3 Android Project Build Process

A simplified version of the Android project build process is depicted in Figure 4-1C, and more details are available at https://developer.android.com/tools/building/index.html.

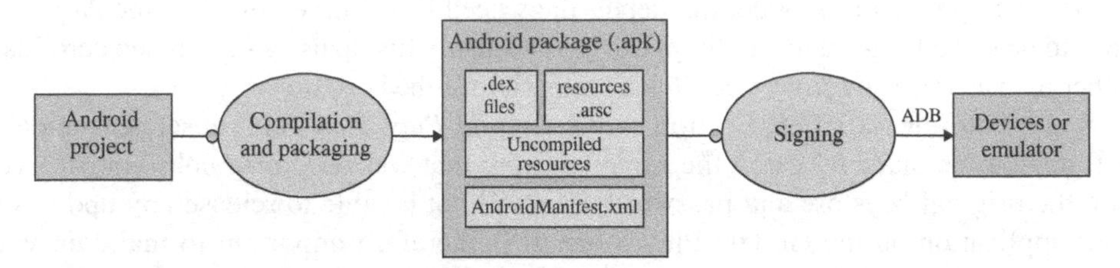

Attribution: Android Developer Documentation at *developer.android.com*.
Figure 4-1C Simplified view of Android application build process.

4.2.4 APK installation and execution

When your App's apk file is installed on an Android device, it is stored under an internal path such as */data/app/** on your device's file system; the exact location is controlled by the Android OS. Due to security constraints implemented by Android, you generally cannot access or even list the installed APK files via the shell environment, using commands such as *adb shell ls*. The quickest way to reliably tell if a package (APK) has been installed is to use the package manager (*pm*) command: *adb shell pm list packages -f*. An installed package corresponds to an APK on a 1:1 basis. The snippet below shows the commands and outputs pertinent to determining if particular packages have been installed:

```
$ adb shell pm list packages -f | grep ted
package:/data/app/com.ted.android-1/base.apk=com.ted.android
$ adb shell pm list packages -f | grep pertino
package:/data/app/com.pertino.connect-1/base.apk=com.pertino.connect
```

When your application is run on a device, it executes within its own isolated, secure sandbox. Android is at the low level an inherently multi-user system and runs each application in a VM process, under a distinct OS user ID that your application is unaware of, but internally the Android OS enforces security and permissions on the basis of this internal OS user ID. This helps keep the private files and data of each application separate and inaccessible from other Apps. The Android OS also ensures that the application accesses resources and functions that are consistent with the permissions that have been explicitly granted to the App by the user. Permissions are granted by the user at install time.

An application runs in a VM within a separate process, and it is isolated from other Apps—it cannot access the private content of other Apps, and in turn, other Apps cannot access its private content. The life cycle of the process is managed by the Android OS depending on the needs of the system.

The Android OS endeavors to enforce the principle of least privilege—Apps can access only the resources and features that are essential for performing work, consistent with the permissions that have been granted by the user and no more.

Although Apps live within their secure sandboxed environment, at the same time, there are mechanisms available that facilitate sharing of data between Apps. It is possible for an application to share data and content securely with another App or even be invoked from another App. Also, two Apps that have been signed by the same keystore/certificate (typically published by the same entity) can share the OS user ID—this is a very deep level of sharing because all files and content of one application are available to the other App, as though the two Apps were the same App.

4.2.4.1 Application main thread / UI thread By default, all the components in an application run in the same OS process and thread called the "main thread" or "user interface (UI) thread." When an application component needs to be started, if a process for the application already exists—because some other component of the same application is running—then Android utilizes that process and its main tread for executing the current component.

Long-running operations should be run in background threads rather than the main thread. In the interest of a good user experience, the Android system enforces application responsiveness by providing the user with an *Application Not Responding* (ANR) dialog whenever an application UI (*Activity*) performs lengthy operations on the main/UI thread and fails to respond to user input for more than 5 seconds. The ANR dialog allows the user to close and terminate such an application. In the case of *BroadcastReceivers* that are background components that engage the main thread of the application in order to handle system broadcast messages, the Android system enforces a 10 second timeout. More details on best practices related to ANR are available at http://developer.android.com/training/articles/perf-anr.html.

4.3 Android SDK API

The Android SDK reference is available at http://developer.android.com/reference/packages.html; Figure 4-2A shows a screenshot of a browser accessing the Android SDK reference documentation.

You will notice that there is an API level setting (toward the top left of the screen) that you can adjust depending on the API level that is of interest to you. API level *21*, which corresponds to Android 5, is relevant for covering the material in this book.

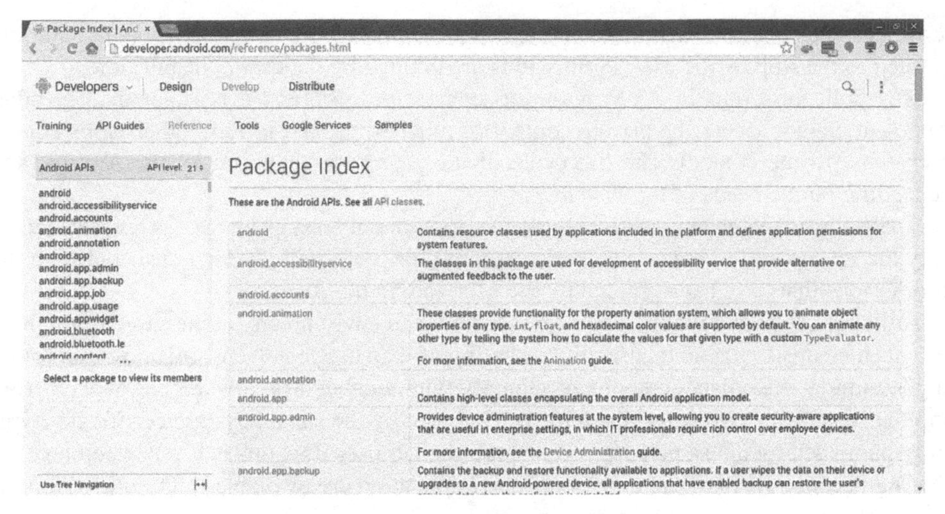

Figure 4-2A Android API reference.

In conjunction with this online API reference, the following resources will be highly useful in order to obtain an understanding of the Android platform:

http://developer.android.com/training/index.html
http://developer.android.com/guide/index.html
http://developer.android.com/guide/components/processes-and-threads.html
http://developer.android.com/guide/components/tasks-and-back-stack.html
http://developer.android.com/guide/components/fundamentals.html
http://developer.android.com/training/basics/data-storage/index.html

4.3.1 Android Application Manifest (*AndroidManifest.xml*)

Every Android application must have a manifest file named *AndroidManifest.xml*. The manifest is a key artifact that contains all the metadata about the application including its package name, the target API level and the minimum API level, security permissions, the application's constituent components, and much more: you may, for example, set your application to be *debuggable* during development or use the *largeHea*p flag to request Android to allocate a larger heap size for running your application. Detailed information on the Android manifest can be found in the Android Developer documentation at:

http://developer.android.com/guide/topics/manifest/manifest-intro.html
http://developer.android.com/guide/topics/manifest/application-element.html

4.3.2 Android API package Overview

At the highest level of the Android package, you will encounter the namespaces *android*, *dalvik*, *java*, *javax*, and *org*. These high-level packages are shown in Figure 4-2B.

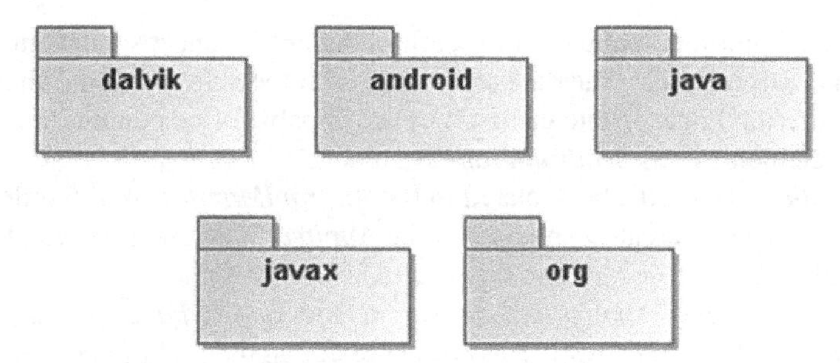

Figure 4-2B Android API high level namespaces, sub-packages.

4.4 Android's Four Fundamental Components

Android application development is based on a few key high-level components that constitute the building blocks of any Android application. These four fundamental components are the *Activity*, *Service*, *BroadcastReceiver*, and *ContentProvider*. Your application components that subclass these four high-level components (or any of their subclasses) must be declared in your Android application project's manifest, that is, *AndroidManifest.xml* file. The life cycle of these components is managed by the Android runtime. This strong Android component model helps separate out the responsibilities of the Application container runtime environment and the application developer.

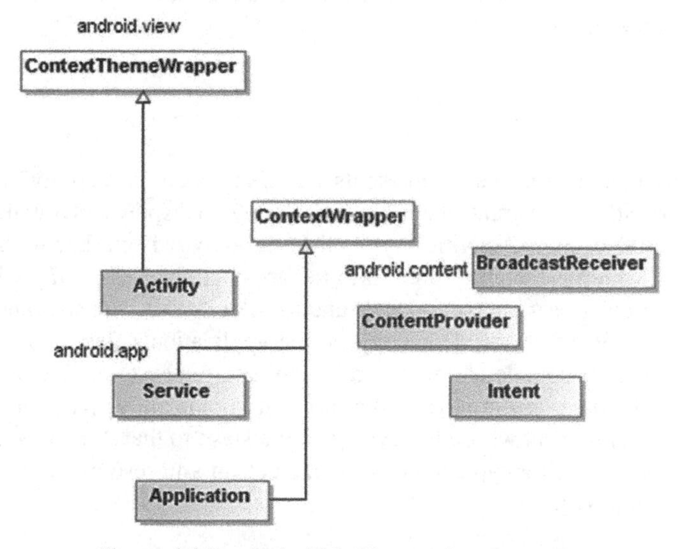

Figure 4-2C Android fundamental components.

Figure 4-2C shows the *Activity*, *Service*, *BroadcastReceiver*, and *ContentProvider* classes that represent the fundamental Android application components. The *Activity* and *Service* classes implement several interfaces that have not been shown in this high-level diagram.

The *Intent* and *Application* classes have also been included in this diagram. The four core components are activated by "Intents." Intents are fundamental to Android application development and runtime application execution. An Intent encapsulates an action to be performed and optionally also the data associated with the action. Intent Filters help components advertise the kinds of intents that they are capable of responding to. You will find "intent-filter" elements in the *AndroidManifest.xml.*

The *Application* class must be declared in the *AndroidManifest.xml* in order to be used. There is often no need to create a subclass of the *Application* class; however, if you choose to do so, it will need to be declared in the *AndroidManifest.xml.*

Activity, *Service*, and *Application* reside in the *android.app* package, while the *BroadcastReceiver*, *ContentProvider*, and *Intent* reside in the *android.content* package.

4.4.1 Android Project Artifacts

An Android project consists of the manifest (*AndroidManifest.xml*), java source files, and resources (res). Resources in turn include layouts, values, xml, raw, assets, and so on. It is recommended that strings be externalized and isolated from the application code. Also, the resource directories can have subdirectories for different form factors and screen resolutions. At runtime, the Android system matches the device's form factor and screen resolution to the appropriate flavor of the resource available in your App. You can and will typically have different flavors of layouts, images, and text that the Android system can pick at runtime, depending on the device's characteristics.

Android resources are an important and vast topic, and you will find useful information in the Android Developer guide at http://developer.android.com/guide/topics/resources/providing-resources.html.

4.5 Activity

The *Activity* is a UI component and represents a user's interaction "activity" with the application. The Android system grants the *Activity* a window to display its UI within. Any given application has one or more Activities; typically, one *Activity* is marked in the manifest as a main *Activity*, which is the entry point into the application. Each *Activity* is intended to implement a particular user interaction and function. An *Activity* can start another *Activity*, and then terminate itself or not, depending on the application's flow and needs. The life cycle of the *Activity* is completely managed by the Android system. The application code should never instantiate an *Activity*; furthermore, it should never pass an *Activity*'s *this* reference to any component whose life cycle is not a slave to that *Activity*'s life cycle. The layout of the *Activity*'s UI is typically externalized as an xml resource file, and is located under the *res/layout* tree:

http://developer.android.com/reference/android/app/Activity.html
http://developer.android.com/training/basics/activity-lifecycle/index.html
http://developer.android.com/guide/components/tasks-and-back-stack.html

Running and previously running Activities are organized as a stack, with one Activity that is in the foreground and interacting with the user, at the top of this *Activity* stack. A task

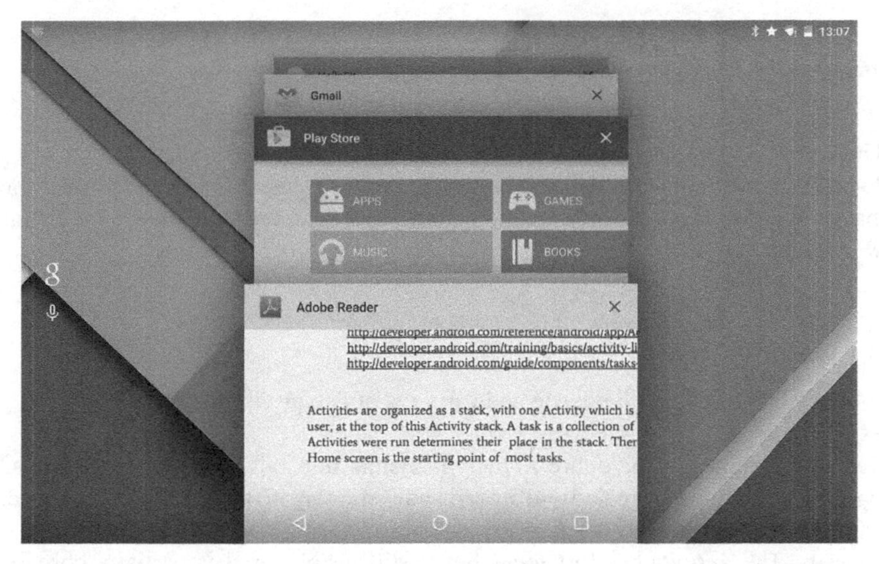

Figure 4-2D *Activity* stack.

is a collection of such *Activities*, and the order in which the Activities were run determines their place in the stack. The device home screen is the starting point of most tasks.

Figure 4-2D shows the *Activity* stack, which you can access by touching the square, recent Apps button on Android 5 devices.

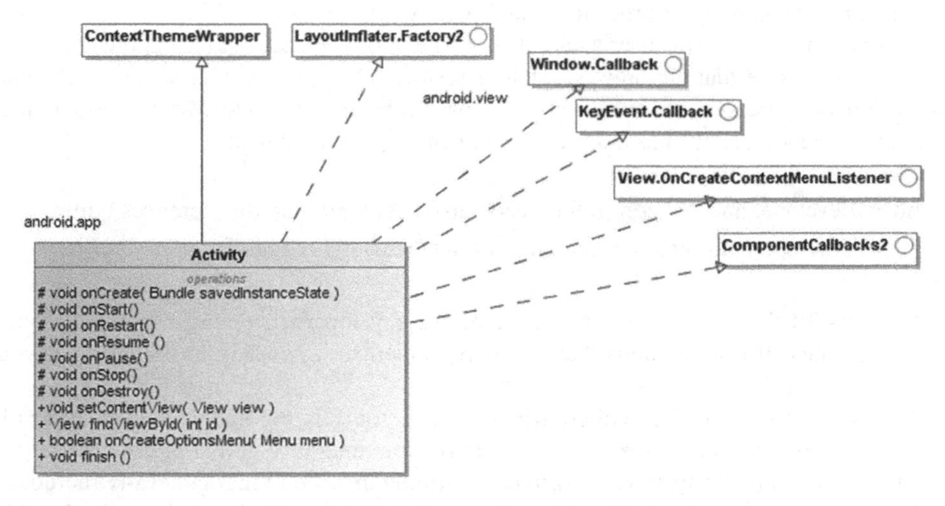

Figure 4-3A *Activity* class, partial listing.

Figure 4-3A shows only a minuscule subset of the operations that are available in the *Activity* class. The *setContentView* method sets the *Activity*'s content to a view that is typically the compiled equivalent of the xml layout—as shown in the snippet below:

```
setContentView(R.layout.main)
```

The call to *setContentView* is typically made in the *onCreate* method, which is a life cycle callback method that is called when an *Activity* is created. The *Activity* life cycle and callbacks are covered in the next section. An *Activity* can terminate itself by calling the *finish* method.

4.5.1 Activity life cycle

The *Activity* has a specific life cycle and can exist at any given time, in one of three given states:

Resumed/Running—The entire *Activity* is visible in a full screen and has focus. The *Activity* has already been created and started and is running in the foreground of the screen. It is at the top of the *Activity* stack.

Paused—The *Activity* has lost focus but is still visible, as it is partially obscured by some other UI. The *Activity* is still alive in that its state and member variables are still intact and it continues to remain attached to the window manager.

Stopped—The *Activity* is no longer visible, as it is completely obscured by some other UI. It continues to retain its state and member variables; however, it is no longer connected to the window manager.

When an *Activity* is in a paused or stopped state, the Android system may kill its process or finish off the *Activity* in order to reclaim OS system resources. When such an *Activity* needs to be displayed to the user again, it will need to be restarted, and the application code will need to ensure that the previous state is restored. *SharedPreferences* is a mechanism via which an application can store key–value pairs related to the state. More information on SharedPreferences, as well as data storage in general, is available at:

http://developer.android.com/reference/android/content/SharedPreferences.html

http://developer.android.com/guide/topics/data/data-storage.html

Figure 4-3B shows a diagram from the Android developer site that shows *Activity* states as well as the callback methods that you may implement according to the needs of your application's *Activity.*

The *onCreate* method is called when your *Activity* is created. You will typically need to override the *onCreate* method. It is recommended to call *super.onCreate* first before implementing any of your *Activity*'s initialization code that generally includes a call to *setContentView* to set the *Activity*'s content to a layout that defines the UI. The onCreate and *onDestroy* methods are the callback hooks into the entire lifetime of the *Activity*. Any global resources initialized in the *onCreate* should ideally be released in the *onDestroy* method.

Similarly, the *onStart* and the *onStop* methods represent the callback hooks into the visible lifetime of the *Activity*. The *Activity*'s visible lifetime includes the foreground

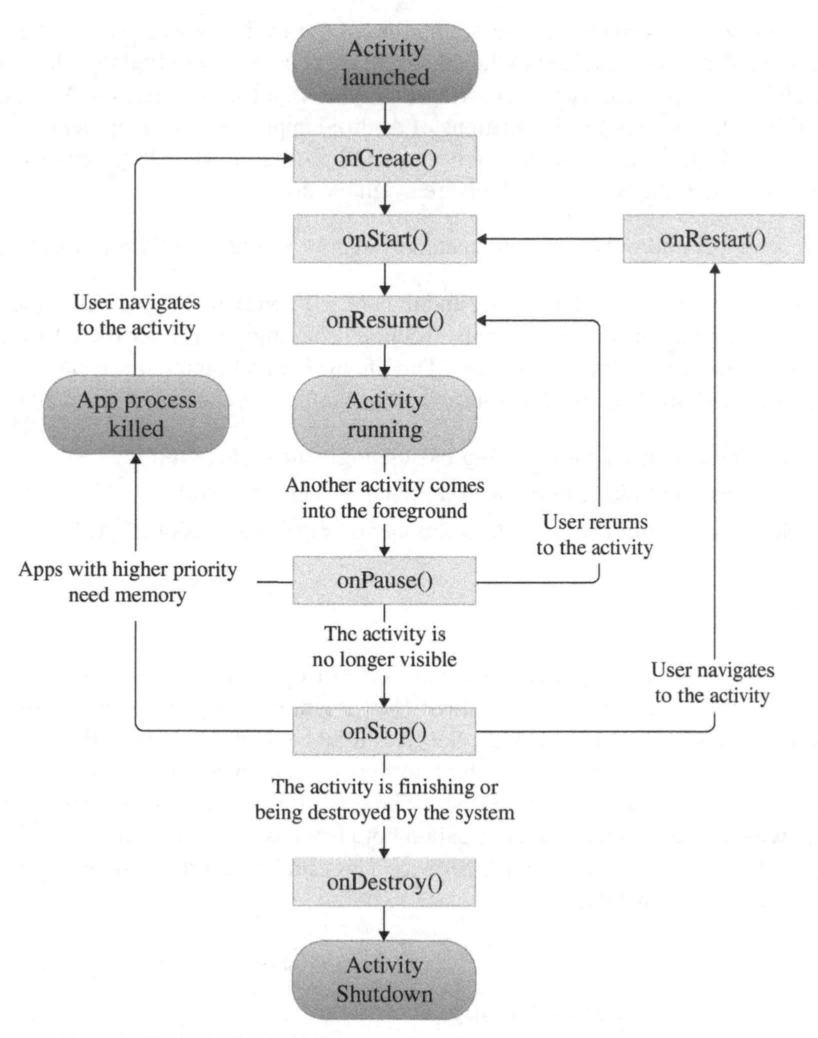

Attribution: Android Developer Documentation at *developer.android.com.*
Figure 4-3B *Activity* life cycle.

lifetime and the partially visible (paused period) of the *Activity*. If you register a listener of some sort in your *onStart* method, you would ideally unregister it in your *onStop* method.

The *onResume* and *onPause* methods represent the foreground lifetime of an *Activity*. Because the *Activity* can go frequently between the paused and resumed states, these methods will tend to be called frequently, and the code in these callback methods is generally lightweight.

One simple, effective way to see your the *Activity's* life cycle callbacks in action is to write a simple *Activity* and include logging in all the callback method implementations. The documentation on the *android.util.Log* class is available at http://developer.android.com/reference/android/util/Log.html. Once you have installed your application to a device, you can watch the logs as you change the orientation of the device or even simply let it "rest" until the screen fades or locks due to the idle timeout. You will likely find that there are a lot more *Activity* life cycle callbacks occurring than you might have expected. Mastering the *Activity* life cycle is one of the foundations of Android application development.

Any Android application runs in one or more OS processes. The OS processes too have their own life cycle, the details of which are available at:

http://developer.android.com/reference/android/app/Activity.html#ProcessLifecycle

While the *Activity* is a widely used fundamental UI component, you will typically use Activities in conjunction with Fragments. Fragments are modular, reusable UI subcomponents that are embedded within Activities. The life cycle of a Fragment is a slave to the life cycle of its containing *Activity*. The following are useful references on Fragments:

http://developer.android.com/training/basics/fragments/index.html
http://developer.android.com/guide/components/fragments.html
http://developer.android.com/guide/components/fragments.html#Lifecycle

4.6 Service

A *Service* is a non-UI component that can perform operations in the background. For example, an *Activity* may start a service to offload potentially long-running operations such as network calls or file I/O to a service. Services may be bound or unbound. In the case of a bound service, other components such as an *Activity* can invoke method calls on the service's interface after binding to the service. The service interface can be implemented via AIDL, in which case the service can be called both from external applications and by components within the same application. There are a cost and an overhead with implementing the service interface via AIDL.

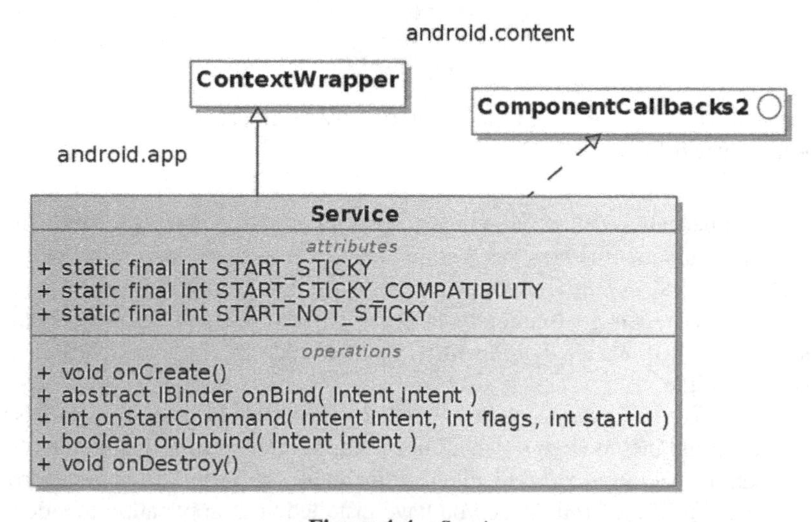

Figure 4-4 *Service.*

Figure 4-4 shows a few of the attributes and methods available in the *Service* class. The *onCreate* method is a callback that is called by the Android system, when the service is first created. The *onStartCommand* is called by the Android system, every time that a component such as an *Activity* invokes the *Context.startService* method using an *Intent* that is associated with the *Service*. The onStartCommand returns to the Android system, an int value that indicates to the Android system how it should manage the behavior of the *Service* if its process gets killed. The constant *START_STICKY* is one of several possible values that are meant to be returned by the *onStartCommand* to indicate to the Android system how the service is to be managed. Particularly, if *START_STICKY* is returned by *onStartCommand* and the process is killed after the service was started, the Android system will try to create the *Service* again. This mode is useful for long-running background Services, such as music playback. If there were any resources that were created or initialized in the onCreate method, the same resources should typically be released in the *onDestroy* callback method.

A *Service* can be started by calling *Context.startService* or *Context.bindService*. The *Context.startService* requires an explicit Intent, starting with Android 5. Use of intent filters is not recommended. Intents and Intent Filters have been covered later in this chapter.

The *Service* class itself is not threaded; nor is it inherently a separate process. By default, it runs in the same process as the rest of the application that it belongs to. The *Service* thus runs in the main thread of its containing process, and you will need to implement a separate thread in the application logic in order to perform long-running work in order to avoid the risk of an ANR situation.

Your service class must be declared in the manifest within the *<service>* element, and the syntax for this is as follows:

The *exported* flag attribute governs whether the service can be invoked by external applications.

```
<service name="com.example.PlayerService"
         android:enabled=["true" | "false"]
         android:exported=["true" | "false"]
        android:icon="drawable resource"
        android:isolatedProcess=["true" | "false"]
        android:label="string resource"
        android:permission="string"
        android:process="string" >
          . . .
</service>
```

Detailed information on services and service element is available at:

http://developer.android.com/guide/components/services.html
http://developer.android.com/guide/topics/manifest/service-element.html
https://developer.android.com/training/run-background-service/create-service.html

4.7 BroadcastReceiver

A *BroadcastReceiver* is a component that can consume a system-wide broadcast message, and it does so via its *onReceive* handler method (Figure 4-5). Your application's broadcast receiver component will need to extend *BroadcastReceiver* and override the *onReceive* method. The nature of the broadcast is asynchronous, and its consumption is executed in the background. However, the Android system considers your application *BroadcastReceiver* instance to be active for the duration of execution of the *onReceive* method, subject to a maximum of 10 seconds.

Using a *LocalBroadcastManager* is preferable, as it entails less overhead and increases security as data will never leave your application when using a local broadcast.

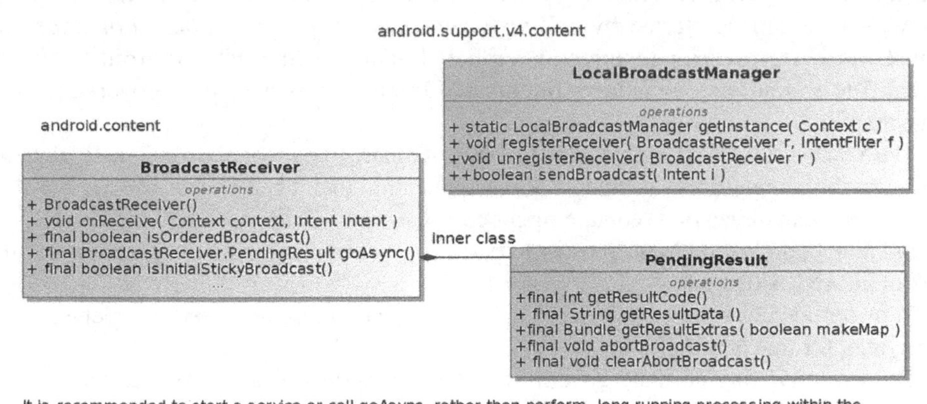

Figure 4-5 *BroadcastReceiver.*

4.8 ContentProvider

ContentProviders are a particular mechanism for storing application data on the Android device (Figure 4-6).

Most Android applications need to persist application data. At the very least, they may need to save some settings or application state so as not to lose data when the application is paused. SharedPreferences are useful for storing key–value pairs, while files can be used to store arbitrary data. Android includes the *SQLite* database for storing structured data and retrieving structured data. *ContentProviders* help abstract out the underlying mechanism of storage and support sharing of data between applications while enforcing security constraints on the data access. The Android API provides off-the-shelf content providers for calendars, contacts, and media files. You may choose to implement your own content provider, especially if sharing of the content is relevant. Other advantages of ContentProviders are that they work well with Sync Adapters (which are useful for keeping content in sync between device storage and cloud-based storage) and search suggestions. More details on

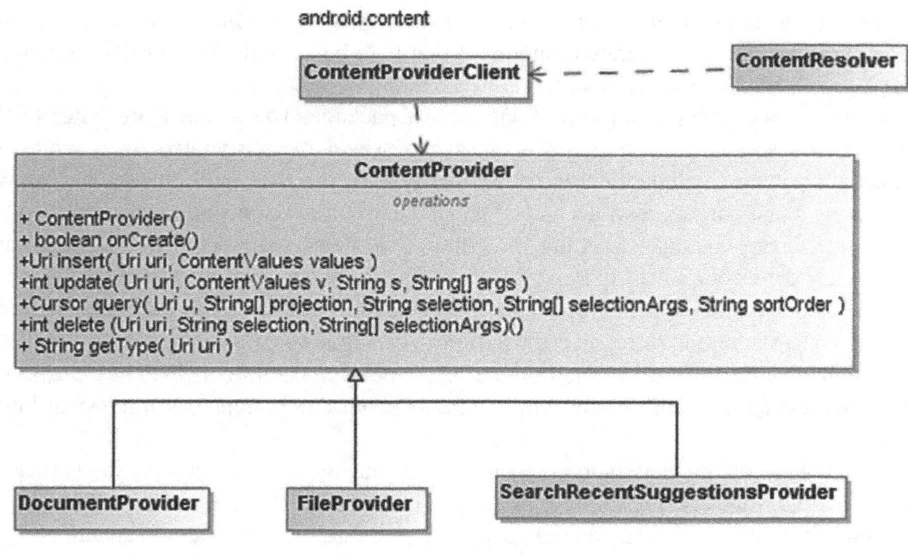

Figure 4-6 ContentProvider.

the general background about Android data persistence, ContentProviders, and Sync Adapters can be found at:

http://developer.android.com/guide/topics/data/data-storage.html
http://developer.android.com/guide/topics/providers/content-providers.html
https://developer.android.com/training/sync-adapters/index.html

4.9 Intent

The Intent is a very powerful and important construct that is fundamental to Android application development and runtime execution. An Intent represents a conceptual definition of an action or operation that needs to be performed. Intents are involved even before the most trivial application can be started on your device.

Although the Intent is merely a passive message or data structure, it is used by the Android platform as a late binding mechanism between different components, which is why an Intent is so powerful. Intents can be used by one component to start another component within the same application or even in a different application. Intents are also used to receive broadcasts either via declaration in the manifest or via the *Context.registerReceiver* method. Intents are intercepted and handled by the Android system at runtime.

In the real world, you may have the intention of lighting up your room, and you may achieve this by flipping on a light switch. There may be more than one switch in the room, or you may even have other mechanisms for turning on the lights (such as via using a smart phone or wearable application to turn a Bluetooth lamp). On the one hand, you have a conceptual action, and on the other hand, you have one or more concrete mechanisms available for invoking that action. Android Intents provide a similar separation of the abstract action from the concrete component that is to be started via "implicit" Intents—which we will be covering shortly.

The *Intent* class resides in the *android.content* package. The primary attributes of the *Intent* are the action (or component's class name) and the associated data, while the secondary attributes include categories, MIME type of the data and extras. The component's class name can act as a secondary attribute when used in conjunction with other attributes. In many instances, setting the component name (fully qualified package and class name) of the component to be started is adequate information for proceeding further. When the action name (*String*) is specified and no component name has been specified, the Android system attempts to resolve the action to a matching component based on all the applications that have been installed on the user's device. The intent filter in the application's manifest declares the Intent actions that a given component has interest in being associated with.

You may have intents that handle custom actions that are not from among the numerous generic actions that are predefined as static constants in the Intent class. The Intent namespace is global so in case you define your own Intent, it is important to ensure uniqueness by using your application's package name as a prefix to the action's name.

The *Intent* class has several constructors including an empty constructor. Such an empty *Intent* will certainly need some attributes to be set—such as an action, component, and possibly data—before it can be meaningfully used. Other available constructors allow you to set the action and/or the component class details in one step while instantiating the Intent. Extras can be set on the Intent using the overloaded *putExtra* method. There are a large number of overloaded versions of the *putExtra* method to cater to different fundamental and commonly used data types.

Figure 4-7 shows some only a subset of the attributes and methods available in the *Intent* class in order to provide a conceptual understanding of the *Intent* class. The complete details of are available in the API reference: http://developer.android.com/reference/android/content/Intent.html.

There are several static final attributes available in the Intent that help define various types of information:

Standard Activity Actions—These are numerous standard actions associated with Activities, and only a few have been shown in the diagram such as *ACTION_VIEW*, *ACTION_DIAL*, and *ACTION_MAIN*. Depending on the particular action, additional data may be required. For instance, *ACTION_DIAL* requires associated phone number data as a uniform resource identifier (URI) (Tel.: 6502752515), which can be parsed using *Uri.parse*. We will be covering standard activity actions in detail shortly, in the section on Implicit Intents.

Standard Broadcast Actions—These are standard actions associated with broadcasts, which *BroadcastReceiver*s can register for, in order to receive broadcasts. The Intent action *ACTION_BOOT_COMPLETED*, for example, represents the completion of the boot process, when your device boots up. In order to implement a *BroadcastReceiver* that has interest in receiving this broadcast intent, you will need

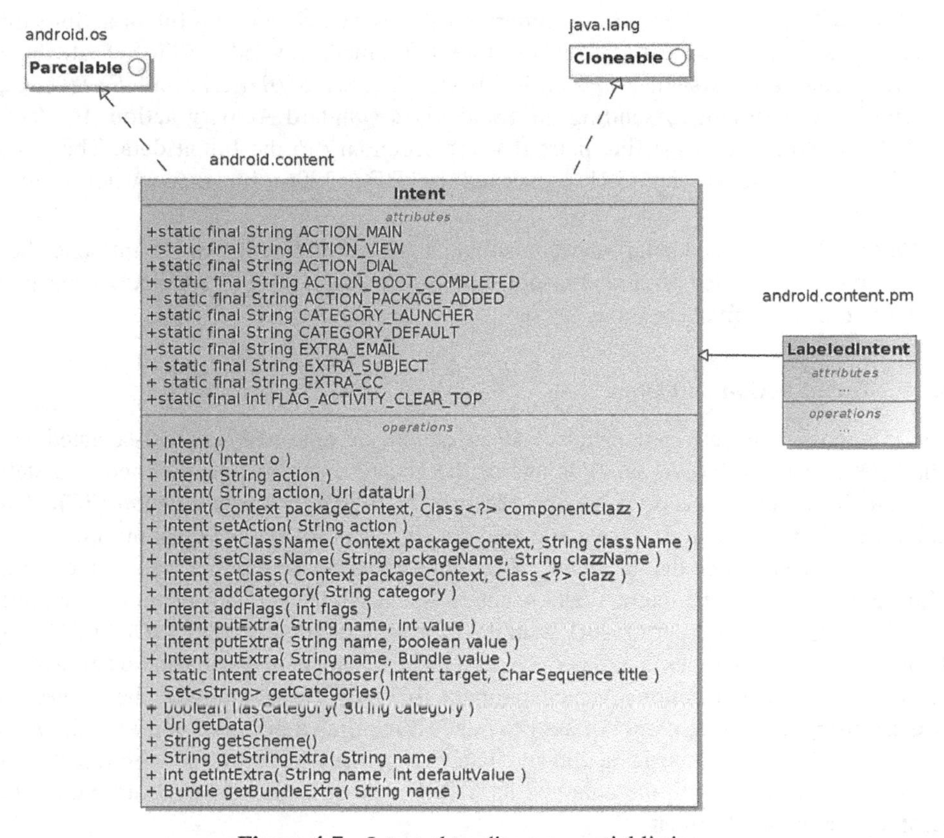

Figure 4-7 Intent class diagram, partial listing.

to declare the *android.permission.RECEIVE_BOOT_COMPLETED* in the application's manifest file.

Standard Categories—These are secondary attributes that can be associated with Intents. *CATEGORY_LAUNCHER* is one such example of a standard category. *CATEGORY_LAUNCHER* is used along with *ACTION_MAIN* in the intent filter of your main *Activity* in your application's manifest—they help declare that the particular (main) *Activity* is the entry point of your application and its icon and label are to be displayed in the application launcher:

```
<intent-filter>
        <action android:name="android.intent.action.MAIN" />
        <category android:name="android.intent.category.LAUNCHER" />
</intent-filter>
```

Standard Extras—These are secondary attributes associated with Intent actions that can be set on and accessed from the Intent instance. Several EXTRA_* attributes have been defined such as *EXTRA_EMAIL*, *EXTRA_SUBJECT*, and *EXTRA_CC*, which are pertinent to sending an email via a standard *Activity* action *ACTION_SEND*. Extras are key–value pairs that are secondary to the Intent data. The Intent data is defined via an URI (governed by RFC 2396; https://www.ietf.org/rfc/rfc2396.txt).

Flags—These are secondary attributes that can be set on Intents, which influence how the Intent is handled by the Android system. Several *FLAG_** attributes have been defined as static final constants.

4.9.1 Intent Action and Data

The Intent as we just covered encapsulates an action and optionally some associated data. The Intent action *ACTION_VIEW* is one of the generic intent actions defined as a static constant in the Intent class. It has the *String* value of *android.intent.action.VIEW*. The Intent *ACTION_VIEW* must have associated data, in order for it to be meaningful.

As mentioned earlier, the *ACTION_DIAL* requires the phone number data, in the format "Tel.: 6502752515." The intent's data needs to be set via the *setData* method. The prefix "Tel.:" is important in order to work with the *Uri.parse* method. The *ACTION_DIAL* only displays the phone number in a dialer program and does not actually make the call without the user's subsequent action. A related intent *ACTION_CALL* initiates the phone call directly using the phone number data provided; it requires a permission *android.permission.CALL_PHONE*. Thus, the action and data go together when working with Intents. The action in conjunction with the data and its MIME type governs how the Android system resolves the (implicit) intent.

4.9.1.1 Intent Extras An Extra is a Bundle of additional information that can be used to deliver information to the component that is being started via the Intent. As mentioned in an earlier section, the *ACTION_SEND* in conjunction with the extras *EXTRA_EMAIL* and *EXTRA_SUBJECT* can be used to send an email. Such standard extras help interoperability across Apps. When you use an Intent to start a component within your own application, you will typically use your application's static final *String* constants "EXTRA_SOMETHING" rather than the standard extras.

4.9.1.2 Intent Flags Intent Flags are secondary Intent attributes that influence the behavior of the component that is called via the Intent. The *FLAG_ACTIVITY_** family of flags pertain to the behavior of Activities that are launched via Intents via the *Context.startActivity* method, while the *FLAG_RECEIVER_** family of flags pertain to the *Context.sendBroadcast* method call.

As an example, if the *FLAG_ACTIVITY_CLEAR_TOP* is set on the *Intent* that is used to start an *Activity*, a new instance is not launched if that *Activity* already happens to be running in the current task; rather, the existing *Activity* will be brought to the forefront and the Intent delivered to it. There are other flags available that facilitate various contrasting flavors of *Activity* behavior.

4.9.2 Explicit Intents

When you create an Intent without providing/setting any action and explicitly set a fully qualified class name of the component that you would like to start or invoke, that is an explicit Intent. The fully qualified class name of the component can be specified via the Intent's constructor parameter or via subsequently calling the setter method on the Intent instance.

Typically, you use explicit intents to start other components within the same App. It is very common for one *Activity* to may start another *Activity* or a *Service* within the same application, and explicit Intents are the ideal way to accomplish this.

The explicit Intent is not a separate class in the API—depending on the specifics of how the Intent is constructed and set up, and particularly if a component or class is set on the Intent instance explicitly while excluding an action, the intent is considered to be an explicit intent.

Explicit intents are more secure and you should always use an explicit intent when starting services. Android 5 (API level 21) enforces this principle by throwing an exception in case you call *bindService* without an associated explicit intent.

4.9.3 Implicit Intents

When you create an Intent and associate it with an action (*String*), typically without specifying the fully qualified class name of the component that you would like to start or invoke, that is an implicit Intent. The action can be specified via the Intent's constructor parameter or via subsequently calling the setter method on the Intent instance.

Implicit Intents are usually used in conjunction with the standard generic actions that have been defined as static constants in the *Intent* class. *ACTION_VIEW* (*android.intent.action.VIEW*) and *ACTION_DIAL* (*android.intent.action.DIAL*) are two such standard actions that are widely used with Implicit Intents.

4.9.4 Intent Filter

The Android system matches an implicit Intent to identify and target the corresponding component that needs to be started. One very widely used intent is the *ACTION_MAIN* (with the category of *CATEGORY_LAUNCHER*), which the Android system uses to find all the main activities of all the Apps installed on the device to populate the application launcher with their names and icons.

You will find that the most trivial of the Android apps that you have written so far including our "Hello Lollipop" series in the last chapter has an intent-filter element in the application manifest (*AndroidManifest.xml*) that resembles the following snippet:

```
<intent-filter>
    <action android:name="android.intent.action.MAIN" />
    <category android:name="android.intent.category.LAUNCHER" />
</intent-filter>
```

Intents and Intent Filters work closely together. The concept of Intent Filters, the intent-filter element in the manifest, and the class *IntentFilter* are topics for further reading at the links provided below:

http://developer.android.com/guide/components/intents-filters.html
http://developer.android.com/guide/topics/manifest/intent-filter-element.html
http://developer.android.com/reference/android/content/IntentFilter.html

4.9.5 Intent Resolution

The Android system matches an implicit Intent to the component (*Activity*) that needs to be started. This process is referred to as Intent Resolution. The Android system filters and resolves intents based on the encapsulated action and the associated data and categories. The data itself consists of the type, scheme, authority, and path.

It is possible that no component matching the implicit intent can be found (based on what Apps the user has installed on their device), and this can cause a crash. In order to avoid the crash, you must first ensure that a matching component exists by calling the method *resolveActivity* on the *Intent*, before calling *startActivity.*

It is also possible that more than one matching component is found. The Android system provides the user with the ability to choose the appropriate application (such as the choice of browsers that match the *ACTION_VIEW* on a Web URL data). The *createChooser* method and the *ACTION_CHOOSER* are associated with providing the calling application the option to display an alternative application/activity chooser with a customized title, upon multiple matching components for the Intent.

4.9.6 Intent Use Cases

The following is a summarization and listing of the ways in which Intents can be used.

4.9.6.1 Starting Activities An *Activity* can be started using an Intent by calling the *startActivity* method available in *Context*. Data and extras may be set on the Intent. In the case of an explicit Intent, the Intent encapsulates the name of the *Activity* class that needs to be started. In the case of an implicit Intent, the Android system resolves the intent based on its encapsulated attributes to a matching component or components.

4.9.6.2 Starting Services A *Service* can be started using an Intent via the *startService* method available in the *Context*. Only explicit Intents are to be used in order to start Services.

4.9.6.3 Delivering Broadcasts Intents can be used to send broadcasts using the *send-Broadcast* method available in the *Context*. Such a broadcast can be sent within the same application of between applications. The *LocalBroadcastManager* is recommended for sending broadcasts within the same process.

4.10 *android* package, sub-packages

The purpose of this section is to provide a bird's-eye view of the Android SDK's high-level packages. The vast majority of the Android SDK's classes and interfaces reside in subpackages under the android namespace. There are two classes that reside directly in the android package, namely, *Manifest* and *R*. The *Manifest* class has a nested class *Manifest.permission* that defines various permissions, and these permissions are universally used via the *AndroidManifest.xml*. The *R* class represents the global resources available to any Android App. Typically, Apps use resources from the R class that is available within their Apps package namespace. Apps can use some resources from the global R class, though drawable resources should be avoided. Resources from the global R class, if used in your application, can cause version dependencies and incompatibilities in your App, so they are best avoided. If you inadvertently import *android.R* into any of your application classes, that can cause the references to your own application's *R* attributes to become unresolvable. More information on the subject of resources is available at:

http://developer.android.com/guide/topics/resources/overview.html

http://developer.android.com/guide/topics/resources/accessing-resources.html

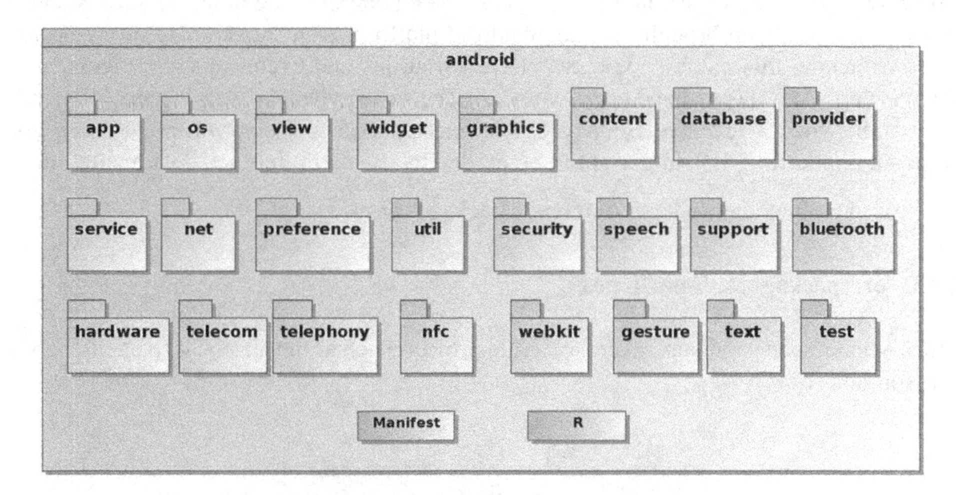

Figure 4-8 android sub-package, partial listing.

Figure 4-8 shows the sub-packages within the android namespace.

4.11 *dalvik* package, sub-packages

The sub-packages and classes under the *dalvik* namespace represent low-level functionality that is seldom used in applications. Yet, they have been included here to emphasize that the deeper system-level functionality is centered around "dalvik."

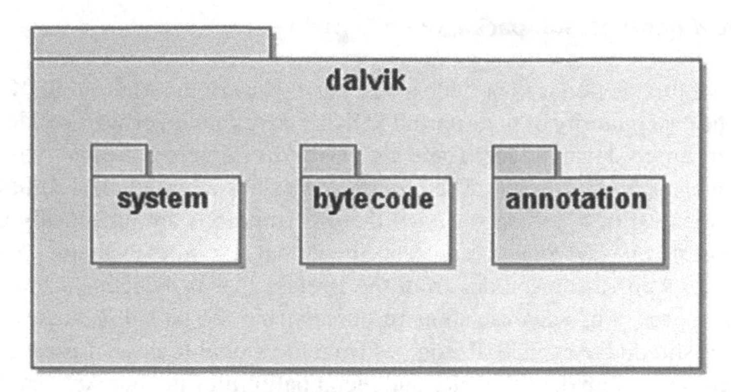

Figure 4-9 dalvik sub-package.

Figure 4-9 shows the sub-packages within the *dalvik* namespace.

4.12 *java* and *javax* package, sub-packages

The *java* sub-packages are familiar to Java developers and resemble the Java Standard Edition that has been brought into the Android platform from Apache Harmony project. The artifacts in this namespace "java" have variations and exclusions—particularly the Java Swing® API classes have been excluded. The security and cryptographic APIs under *java.** and *javax.** leverage Bouncy Castle and OpenSSL. When you are writing your Android application, it is important to be guided by the API reference documentation at

http://developer.android.com/reference/packages.html.

4.13 *org* package, sub-packages

The *org* namespace contains several external projects such as the apache http client, parsers for json, and xml.

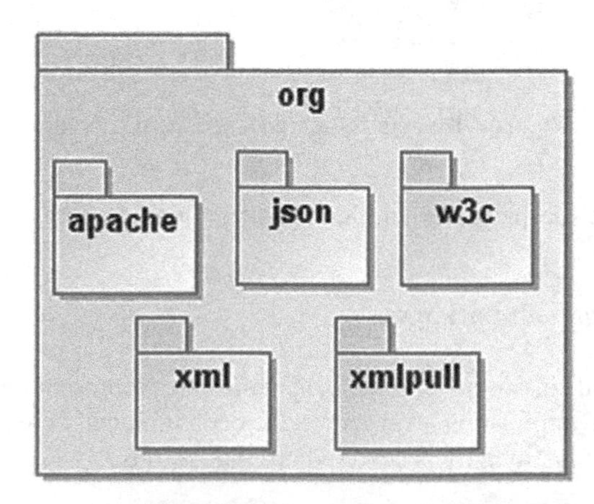

Figure 4-10 org sub-package.

Figure 4-10 shows the sub-packages such as *apache*, *json*, *xml*, and so on.

4.14 Sample code in this book

The source code for this book is organized into three top level directories:

lollipop
wear
fit

The source code under *lollipop* covers the base Android platform and has the following project subdirectories:

0hello1 1hello1 2hellostudio 3ui 4alarm 5services 6provider 7maps 8notifications
Each project has a README file with relevant build instructions.

References and Further Reading

http://en.wikipedia.org/wiki/Component-based_software_engineering#Software_component
http://en.wikipedia.org/wiki/Java_servlet
http://en.wikipedia.org/wiki/Common_Object_Request_Broker_Architecture
http://www.oracle.com/technetwork/java/javase/tech/index-jsp-138781.html
http://en.wikipedia.org/wiki/JavaBeans
https://source.android.com/devices/tech/dalvik/dex-format.html
http://en.wikipedia.org/wiki/Internet_media_type
https://developer.android.com/tools/building/index.html
http://developer.android.com/tools/help/proguard.html
http://developer.android.com/tools/publishing/app-signing.html
http://proguard.sourceforge.net/
http://developer.android.com/tools/publishing/app-signing.html
http://developer.android.com/reference/packages.html
http://developer.android.com/training/index.html
http://developer.android.com/training/articles/perf-anr.html
http://en.wikipedia.org/wiki/Bouncy_Castle_%28cryptography%29
https://www.bouncycastle.org/
http://en.wikipedia.org/wiki/OpenSSL
https://www.ietf.org/rfc/rfc2396.txt
http://developer.android.com/guide/components/services.html

Chapter 5 Android Device Discovery and Communication

5.1 Android Interconnectivity

Inter-device communication has much significance in a world of a multitude of devices and peripherals that reside in the body area and home area networks. In such a world of a multitude of devices that a consumer may possess or interact with, it is important that devices can interconnect conveniently and securely. The specifications and standards that address this important function are by no means unique to Android; rather, the base Android APIs typically support common industry standards that boost such interconnectivity between devices and services.

As users, we have probably seen our devices interact with other devices via various technologies, such as *Near-Field Communication (NFC)*, *Bluetooth*, *Wi-Fi*, *Wi-Fi* Direct, etc. As developers, we have written or will end up writing at some point, applications that leverage these technologies.

Consumers typically find it cumbersome to have to set up network-based printers, scanners, TVs, etc. via configuration of technical attributes such as IP addresses, ports, and so on. Consumers tend to appreciate it when the devices and their associated applications are self-configuring and can interconnect with minimal manual steps.

As technologists, it behooves us to make technology easy for general consumers to use by reducing the manual and cumbersome steps that can potentially act as barriers to adoption and usage.

Wearable Android™: Android Wear & Google Fit App Development, First Edition. Sanjay M. Mishra.
© 2015 John Wiley & Sons, Inc. Published 2015 by John Wiley & Sons, Inc.

5.2 Advertisement and Discovery

Advertisement and discovery are important core concepts that go together; they help simplify and automate the process for devices to interconnect and work together from the consumer's perspective. Associated with advertisement and discovery are the underlying operations such as scanning, detecting, filtering, pairing, and tethering. The concept of advertisement and discovery exists in some form or the other in several of the connectivity standards and technologies.

5.3 Bluetooth

Bluetooth (classic) represents a family of wireless specifications and protocols that address the exchange of data between wireless devices over short distances, in the order of a few feet. The Bluetooth standards are managed by the Bluetooth Special Interest Group. Bluetooth supports *advertisement* of the devices and the services that they offer. A service is identified by a standard 128-bit universally unique identifier (UUID). This makes it possible for devices to advertise their presence including details such as their name and the services that they offer. Bluetooth also supports *service discovery*, which allows a device to discover other devices and the services that they offer. Bluetooth supports various protocols such as Point-to-Point Protocol, TCP/IP and UDP, Object Exchange Protocol, and *Wireless Application Protocol*.

5.3.1 Bluetooth Low Energy (LE)

Bluetooth "Low Energy" (LE)—also known as *Bluetooth Smart*—is a standard also managed by the Bluetooth Special Interest Group. Bluetooth LE aims to reduce power consumption and makes it easier for smaller devices such as fitness sensors to communicate with other devices. As it turns out, Bluetooth LE is not backward compatible with classic Bluetooth. Detailed information on Android's support for Bluetooth and Bluetooth LE can be found at:

http://developer.android.com/guide/topics/connectivity/bluetooth.html
http://developer.android.com/guide/topics/connectivity/bluetooth-le.html

In order for your application to connect to a Bluetooth device, it will need the *android.permission.BLUETOOTH* permission. In order for your application to discover and pair with Bluetooth devices, it will require the *android.permission.BLUETOOTH_ADMIN* permission.

5.3.2 Bluetooth Generic Attribute Profiles (GATT)

Bluetooth Attribute Profile (ATT) is a wire application protocol, closely associated with Generic Attribute Profile (GATT). GATT profiles are standard definitions of data and services. More information about GATT is available at:

https://developer.bluetooth.org/gatt/profiles/Pages/ProfilesHome.aspx

In late 2014, a new version of the Bluetooth standard, version 4.2, was announced—which improves upon *Bluetooth LE* in several aspects—including IPV6 support, direct Internet connection capability and faster speed. *Bluetooth 4.2* offers other useful privacy features including consumer controls to make a Bluetooth device invisible.

5.3.3 Android support for Bluetooth LE

The base Android API addresses the pairing and tethering of *Android Wear* devices with Android handheld devices over Bluetooth. The *Google Fit* API—covered in Chapters 8 and 9—has independent, self-contained support for fitness sensor devices that support *Bluetooth LE* and standard GATT profiles. *Google Fit* inherently supports only *Bluetooth LE* devices. It is possible for *Google Fit* to work with devices that support other connectivity technologies—*Google Fit* provides the building blocks for you to create your own software-based sensor and make it available to the *Google Fit* platform. After that, the *Google Fit* platform can interoperate with your custom software sensor via standard *Google Fit* interfaces, as though it were a regular hardware sensor.

5.4 Wi-Fi Peer-to-Peer (Wi-Fi Direct)

Wi-Fi is a widely used technology, which addresses the interconnectivity needs of local devices, typically in a home or office. Various *Wi-Fi*-enabled devices such as personal computers, handheld devices, cameras, game consoles, and so on connect to a wireless access point (cum router), which in turn often provides access to the Internet. *Wi-Fi* is already a widely used and extremely popular technology. The *Wi-Fi Alliance* owns the trademark and manages the *Wi-Fi* standard.

Wi-Fi Direct, also known as Wi-Fi P2P, is a peer-to-peer-based *Wi-Fi* standard extension (also managed by the *Wi-Fi Alliance*) that addresses the subject of direct connectivity between two *Wi-Fi* devices, without the need for a intermediate *Wi-Fi* access point. At least one of the devices will need to support *Wi-Fi Direct* in order for connectivity to work.

Devices supporting *Wi-Fi Direct* typically implement and embed a software-based wireless access point and eliminate the need for the physical hardware-based wireless access point, which we are typically accustomed to using. Such a software-based *Wi-Fi* Direct access point can be simple or even quite sophisticated; it can act as a router and serve as a bridge to the Internet.

Just one sophisticated Wi-Fi Direct-enabled device in a local network can provide connectivity for several legacy *Wi-Fi*-"only" devices and provide them all with Internet connectivity as well.

Wi-Fi Direct works over *Wi-Fi*, which has a better range and faster speed than several other interconnectivity technologies.

Compared to Bluetooth (classic), *Wi-Fi Direct* offers longer connectivity range and faster speeds. Compared to Bluetooth LE, *Wi-Fi Direct* offers a better range. Compared to NFC technology covered later in this chapter, *Wi-Fi* Direct offers a better data transfer rate.

Wi-Fi Direct's ability to bridge the path to the Internet for other devices tends to reduce the need for individual smaller devices to possess independent Internet access. This can be very useful in a world of a multitude of devices including wearables and IoT devices.

5.4.1 Android Wi-Fi Direct / P2P API

Android's *Wi-Fi Direct/P2P* API allows Android devices with the appropriate *Wi-Fi Direct* hardware to connect directly to each other over Wi-Fi, without an intermediate access point. This API helps your application in discovering, requesting, and connecting to peers without being connected to the conventional network. Many devices and peripherals such as cameras, projectors, printers, scanners, and sensors have support for *Wi-Fi Direct*. More information about Android's support for *Wi-Fi Direct* can be found at:

http://developer.android.com/guide/topics/connectivity/wifip2p.html
http://developer.android.com/training/connect-devices-wirelessly/nsd-wifi-direct.html

5.5 Zero Configuration Networking (zeroconf)

Devices—much like humans in society—can be organized to effectively "meet and greet" other "peers" devices on the local network and share their (host/device) names and their "occupations": the particular services they offer (provide). *Zero configuration networking* (*zeroconf*) is, today, a set of open standards and technologies that makes it easier for devices and applications to interconnect and interoperate over the local TCP/IP network, with minimal manual setup and configuration—hence the term "zero configuration."

Stuart Cheshire who holds a Ph.D. from Stanford University pioneered the ideas behind *zero configuration networking* and led its development since the mid-1990s. Working at Apple, he authored several IETF RFCs that aim to make this technology an open standard. Cheshire, along with Daniel Steinberg, has coauthored a book titled *Zero Configuration Networking: The Definitive Guide* published by O'Reilly.

Various *zeroconf* software protocols aim to make it possible for devices and applications to advertise their existence on the network and discover other services on the network. Service discovery is the concept and set of protocols that are a part of *zeroconf*, which address the automatic discovery of services on the network. Ultimately, *zeroconf* makes it easier for consumers to get their devices and applications to interoperate and work together, without cumbersome manual steps.

Zeroconf is built on top of core technologies such as the assignment of network IP addresses to networked hosts (devices) on the network, automatic distribution and resolution of hostnames, and location of network-based services. Multicast Domain Name Service (mDNS) is used to resolve hostnames to IP addresses within small networks.

Zeroconf is aimed at local consumer networks, which typically have a limited number of hosts. *Zeroconf* was originally released by Apple as Rendezvous, but it was later renamed to Bonjour®. *Zeroconf* is platform and vendor agnostic and interoperates across OS platforms. *Avahi* is an open-source GPL-based implementation of *zeroconf* for Linux; implementations on most major platforms are also available.

Universal Plug and Play (*UPnP*) is another standard and set of protocols that supports *zeroconf* and addresses detection and discovery of consumer and entertainment devices on the network. *UPnP AV* is an extension that addresses the particular needs of audio and video devices such as TVs, VCRs, CD/DVD players, media servers, set-top boxes, stereo systems, and so on. *UPnP* is also vendor and platform independent. *UPnP*-compatible devices are able to advertise their presence and capabilities on the network, so that other devices and applications can discover them and avail of their services over the network.

Since several years, it has become common for consumer devices such as printers, scanners, and TVs to support *zeroconf* and/or UPnP. As more and more consumer devices and home appliances including washers and dryers, light bulbs, and so on commence to join the local IP network, the relevance and potential for *zeroconf* will tend to increase.

While some health- and fitness-related sensor devices are typically carried around by the user, others such as a weigh scale, for instance, tend to "reside" within the consumer's home. It becomes relevant for a weigh scale to support some form of connectivity such as Wi-Fi or Bluetooth, so that the fitness readings (besides being displayed via the weight scale's display) can also be available to other applications on devices such as phones, tablets, netbooks, and so on. Wi-Fi has a better range than Bluetooth, and Wi-Fi-based connectivity in conjunction with *zeroconf* can make it easier for the weigh scale's network service and readings to be discoverable and accessible over the local IP network.

Zeroconf is thus relevant both for some health-related sensors and home automation integration from mobile devices, including phones and smart watches, over the local wired and wireless IP network.

5.5.1 Android Network Service Discovery (NSD)

Android has support for Network Service Discovery (NSD) which helps your application detect other devices and services on the local network, which can be useful for file sharing, gaming, and other applications including home automation and fitness sensor devices that are part of the consumer's local network. Details of Android's support for NSD can be found at:

http://developer.android.com/training/connect-devices-wirelessly/nsd.html

5.6 Near Field Communication (NFC)

Near Field Communications (NFC) is a technology that operates over a short range of about 4 centimeters (cm) or less to initiate a connection between two devices. NFC is suitable for sharing small payloads such as a visiting card, URLs, credit card payment information, and so on. Compared to classic Bluetooth, NFC has a shorter range, consumes less power, and operates at slower speeds. Some health and fitness sensors use NFC-based connectivity.

NFC has no concept of pairing between devices. NFC communication involves an initiator and a target; the initiator creates the radio-frequency (RF) field that can power a passive target called the NFC tag. NFC tags can store data up to 4096 bytes and are typically "read only," but they can also be rewritable. NFC tags are used in key fobs and cards.

NFC can be used on a peer-to-peer communication basis when both sides are powered—such as two phones. Android Beam®—a feature built into Android—simplifies peer-to-peer communication between two Android devices by initiating communication via tapping the phones together. The connection is automatically started when two devices are in range and requires user actions prior to data transmission and exchange.

More information on Android's support for NFC can be found at:

https://developer.android.com/guide/topics/connectivity/nfc/index.html

5.7 Universal Serial Bus (USB)

Universal Serial Bus (USB) is a technical standard and protocol that addresses interconnectivity between a host device and a peripheral device, via direct physical or wired connections. A wide variety of computer peripherals and devices such as keyboards, mouses, printers, scanners, cameras, health monitors, game consoles, mobile devices, and so on use USB standards for connectivity. Some health monitors and sensors use USB-based connectivity.

USB was developed in the mid-1990s and has been in use for over a decade; there have been many updates to the versions of the specifications over the years, the most recent—at the time of writing—being version 3.1. The USB interface addresses both data communication and the supply of power. The USB host device initiates all communications, while the USB peripheral/accessory device responds to queries from the host. The USB host can supply power to the USB peripheral/accessory.

5.7.1 USB On-The-Go (USB OTG)

USB On-The-Go (USB OTG) is an extension that is part of the USB 2.0 standard that allows two devices to negotiate which of them will play the role of the USB host. USB OTG is very useful because a device can act as a peripheral to a larger "host" device at one time and also act as a host to a typically smaller device at another time. For instance, an Android phone can serve as a host in USB host mode to an accessory such as a USB speaker or microphone and power the USB bus to supply power to such an accessory. The same Android device can serve as a storage peripheral when connected to personal computer in another situation and get its battery recharged over the USB during this period.

More information on Android's support for USB host and accessory/peripherals can be found at: https://developer.android.com/guide/topics/connectivity/usb/index.html

References and Further Reading

https://www.bluetooth.org

http://en.wikipedia.org/wiki/Bluetooth_Special_Interest_Group

http://developer.android.com/guide/topics/connectivity/bluetooth.html

https://developer.android.com/guide/topics/connectivity/bluetooth-le.html

http://www.bluetooth.com/SiteCollectionDocuments/4-2/bluetooth4-2.aspx
http://www.wi-fi.org
http://www.wi-fi.org/discover-wi-fi/wi-fi-direct
http://en.wikipedia.org/wiki/Zero-configuration_networking
http://en.wikipedia.org/wiki/Stuart_Cheshire
http://en.wikipedia.org/wiki/Simple_Service_Discovery_Protocol
http://en.wikipedia.org/wiki/Universal_Plug_and_Play
http://en.wikipedia.org/wiki/Near_field_communication
https://developer.android.com/guide/topics/connectivity/nfc/index.html
http://en.wikipedia.org/wiki/USB
http://www.usb.org
http://en.wikipedia.org/wiki/USB
http://en.wikipedia.org/wiki/USB_On-The-Go
http://www.usb.org/developers/docs/

Part III **Android Wear Platform and SDK**

This section covers the ***Android Wear*** platform and *API*, as well as the hands on steps of connecting and setting up of Android Wear devices for development and debugging.

Wearable Android™: Android Wear & Google Fit App Development, First Edition. Sanjay M. Mishra.
© 2015 John Wiley & Sons, Inc. Published 2015 by John Wiley & Sons, Inc.

Chapter 6 Android Wear Platform

6.1 Android Wear

Android Wear is the official Wearable flavor of the **Android OS** from Google, Inc. designed for smart watches and similar devices. *Android Wear* was formally announced in March 2014 and released soon after. Many major hardware vendors, chip makers, and jewelry designers announced their support, and some of them released *Android Wear* devices to the consumer market shortly thereafter. The numerous *Android Wear* partners include Samsung, Motorola, LG, HTC, ASUS, Broadcom, Qualcomm, Intel, Fossil, Inc., MediaTek, and Imagination Technologies, and the list is growing.

The open nature of the Android platform makes it possible for hobbyists, enthusiasts, and businesses to port and extend the Android OS to a wider range of devices over and above those supported officially. There are innumerable such efforts and one such endeavor was the **WIMM One** smart watch from **WIMM Labs**, which was released in 2011.

The WIMM One was based on the Android OS version 2.1 code base. Figure 6-1 shows a WIMM One watch from WIMM Labs, circa 2011. Google, Inc. acquired WIMM Labs sometime in 2012. Thus, the *Android Wear* platform appears to have its origins in the WIMM One smart watch.

Wearable Android™: Android Wear & Google Fit App Development, First Edition. Sanjay M. Mishra.
© 2015 John Wiley & Sons, Inc. Published 2015 by John Wiley & Sons, Inc.

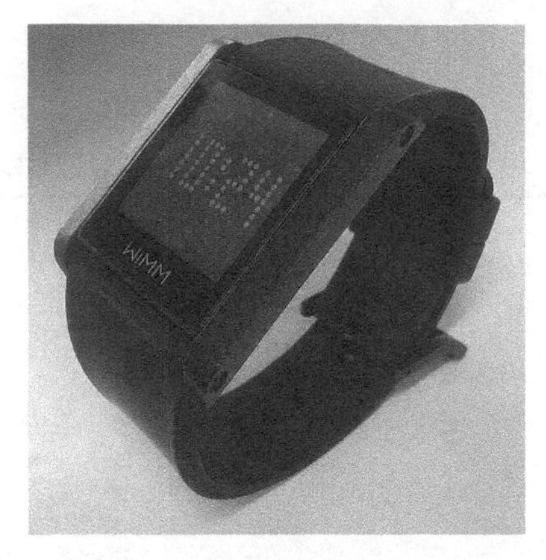

Attribution: Wikipedia user Bostwickenator, under Creative Commons Sharealike license 3.0.
Figure 6-1 WIMM One Android-based smart watch from WIMM Labs, circa 2011.

6.2 Android Wear Platform: Android Wear OS, Wear Devices, and Wear API

The *Android Wear* platform consists of the Wear OS, the Wear devices, and the APIs that support writing Wear Apps. Wear devices are available in a variety of models from numerous manufacturers.

6.2.1 Android Wear OS

The *Android Wear* OS, at the time of writing, runs on ARM and Intel x86 (Atom) family CPUs. The *Android Wear* OS is constrained to be lightweight and subsets the standard Android OS and application runtime environment—meaning that a few of the standard Android APIs are not available on *Android Wear.*

6.2.2 Android Wear Devices

Android Wear devices run a full-fledged OS—the *Android Wear* OS. Wear devices have the ability to execute Wear Apps directly on them. *Android Wear* devices have two standard shapes: square or round.

Figure 6-2A Square Android Wear device, booting up.

Figure 6-2B Square Android Wear device, showing Wear logo while booting up.

Figures 6-2A and 6-2B show a square *Android Wear* device displaying boot animation and the *Android Wear* logo while booting up. *Android Wear* emphasizes simple computer–human interaction and interfaces, based on touch and voice. *Android Wear* supports voice-based interactions. "OK Google" is the "hot word" that triggers voice-based interaction. It tells your device that you are ready to issue a voice-based command. There are several system commands such as "Take a Note," "Set an Alarm," and "Take a Picture." You can also perform a voice-based search using your *Android Wear* device.

Due to the small form factor and limited real estate, interactions are constrained to be extremely simple, requiring minimal human input. They are also intended to be minimally intrusive to the user's attention. Additionally, *Android Wear* aims to align with fitness, workout, and health metrics.

6.2.3 Android Wear API and Wear Apps

Android Wear Apps need to be targeted specifically for and run exclusively on the *Android Wear* platform (API level 20). *Wear* Apps have the ability to access the hardware, sensors, GPU, and services available on the Wear device. *Android Wear* Apps can access a large subset of the standard Android API. Particularly, a few packages are unavailable to Wear Apps, as listed below:

android.webkit
android.appwidget
android.print
android.app.backup
android.hardware.usb

This may at first seem like a limitation, but fortunately, Wear devices are typically tethered to handheld Android devices, and since all Wear Apps released via the Google Play Store have a corresponding "companion" App on the handheld device, some limitations can often be bridged by leveraging the companion handheld App.

6.3 Android Notifications and Android Wear

Notifications were introduced minimally in Android version 3 (Honeycomb) and have had major upgrades in both Android 4.1 (Jellybean) and Android version 5 (Lollipop). Notifications are now a very important part of the Android platform and aligned with the Suggest paradigm. There is a close association between Notifications and *Android Wear*.

Notifications are a mechanism by which Apps that are not currently in the foreground can inform the user that some new information has arrived or that some event has occurred. Notifications are managed by the Android OS and the system UI. An App makes the request to the Android platform to display a Notification, and the Android OS issues the Notification that the user sees in the system-wide Notification status bar. In order to see details of the Notification, the user needs to open the Notification drawer. Both the Notification area and the Notification drawer are managed by the system. After a user receives a Notification, it can also serve as an entry point directly into an App's user interfaces, typically to display finer details about the new information and offer actions on that information via the App's user interfaces. This makes Notifications a strategic entry point into your App and an effective mechanism for user engagement—if leveraged properly.

Notifications include timely reminders and suggestions based on the user's current context. Android users are familiar with the Notification area in the status bar, which is typically located on the top left of their device's screen. A missed call, a newly arrived email, and an upcoming calendar appointment are some examples of Notifications that users are accustomed to seeing. The Notification system allows users to keep up to date with information that they are likely to find of interest. Notifications are closely related to the Suggest model of user interaction. The user may optionally perform actions via the user interfaces presented by the Notification's details; therefore, the latter and optional aspect of the Notification flow can take the user into the Demand model of interaction.

Notifications have various Priority flags in the API—such as default, minimum, and maximum—that the originating App can set in the application code. The use of the Priority *PRIORITY_MIN* flag results in the system UI showing the Notification within the expanded Notification drawer only whenever the user happens to open it. This approach is an "opportunistic" and less intrusive flavor of Notifications—wherein nothing is displayed in the status bar and information is displayed only within the Notification drawer, whenever the user happens to opens it. In addition to Priority, a Notification has several attributes such as a message, an associated icon, an expanded message, and also optionally an action that takes the user into the App's user interface. Other attributes include flags for indicating whether the Notification is to be on a onetime alert basis or as an ongoing event.

6.3.1 Android 5.0 (Lollipop) Notifications

The arrival of *Android Wear* into the Android ecosystem coincides approximately with the release of the Android 5.0 (Lollipop) platform, which has added new features that help the user control Notifications and Interruptions. One of the goals of *Android Wear* is to enable the user to engage in their real-world activities better and reduce the overheads and Interruptions associated with keeping up with the online world. Coincidentally, the core Android platform itself has embraced the principle of giving users more control over Notifications and Interruptions. Notifications have undergone significant changes,

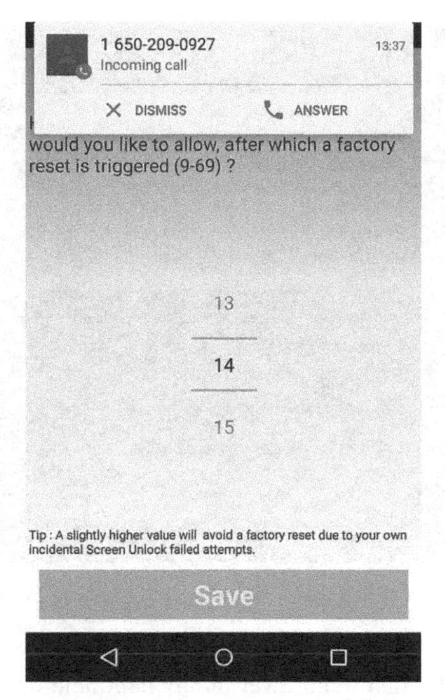

Figure 6-3 Heads-up Notification.

and these changes in functionality, user interface, and structure of Notifications represent somewhat of a departure in design with respect to earlier versions of Android.

Notifications have become more accessible and configurable. The user can choose to receive Notifications through the device screen lock, with control over whether sensitive content may or may not be displayed through the locked screen. **Heads-up** Notification is a new format for receiving high-priority Notifications via a small floating window while some other Apps are in active use.

Users can receive, dismiss, or act on a heads-up Notification without leaving the App they were using. You have likely noticed the heads-up Notification's small floating window, when receiving an incoming call while using any random App on your Android 5 device. Figure 6-3 shows such a heads-up Notification about an incoming phone call, which can be handled without having to leave the App that you are using.

Cloud-synced Notifications are a mechanism for dismissing a Notification on all devices after the user has dismissed the Notification on one device.

The visual design has undergone changes consistent with the new material design theme.

Android 5.0 Notifications also introduce Wearable extensions for *Android Wear*, which we will be covering shortly. You will also notice that Notifications for ongoing events and running services are displayed persistently in the Notification drawer for the duration of the event. It is very useful for the user to remain aware of both the ongoing events and running services on their devices, such as an ongoing phone call.

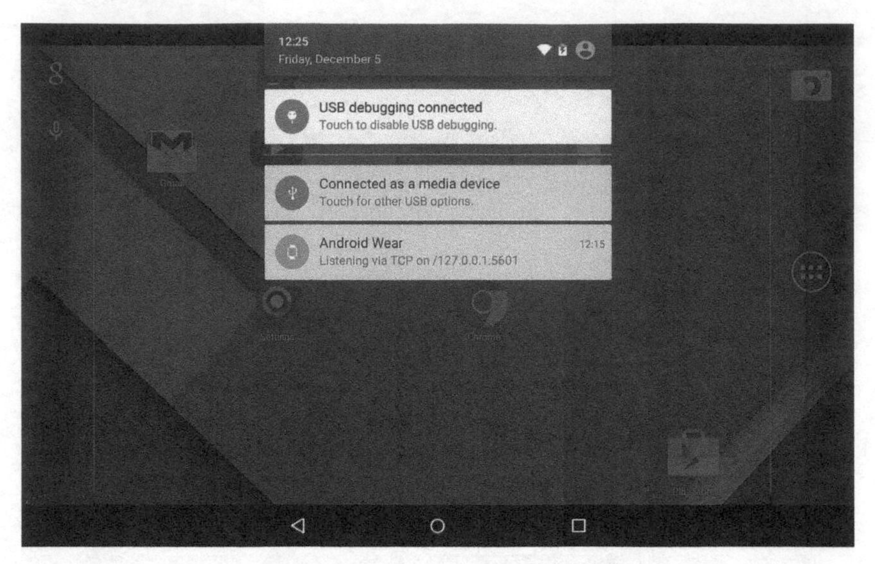

Figure 6-4 Android 5 Notification drawer showing ongoing events or running services.

Figure 6-4 shows a Notification drawer on my handheld device with several ongoing Notifications such as USB debugging and an *Android Wear* service listening on a TCP port.

6.4 Notification Settings and Control

The intricate configurations and control that users have over Notifications affect the behavior of Apps on their handheld devices as well as on the tethered Wear devices. Most Notifications that are shown on your handheld device will also be shown on your tethered Wear device. When you dismiss a Notification on your watch, it is also dismissed from your handheld device. Your handheld device and Wear device can be "muted" independently. It is possible for your Wear device to receive Notifications while your phone is "muted," depending on the per App *Notification Settings* that are configurable by the user.

6.4.1 Sound and Notification and Priority Notification

You have likely noticed that *Sound and Notifications* have been grouped together in the OS *Settings*. You will also have noticed that touching the volume control button on your handheld Android device shows a *Sound and Notification* control widget.

Figure 6-5 shows the Sound and Notification quick control widget with the default settings. The volume control slider controls the volume for Notification alerts. The volume slider responds both to the hardware volume +/– control buttons as well as to touch. The Notification control includes the Notification level options: NONE, PRIORITY, and ALL. The default setting is ALL (seen in Figure 6-5), which results in the Android OS providing

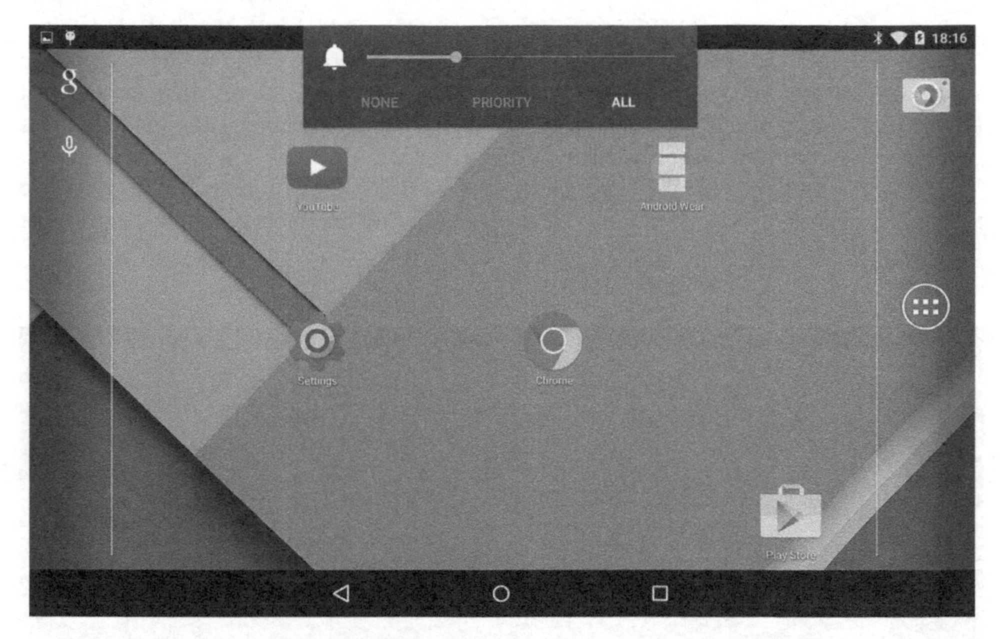

Figure 6-5 Sound and Notification, quick control widget—default Notification level ALL.

ALL Notifications to the user while applying the volume level per the setting in the slider. The volume control works in conjunction with, but is secondary to, the Notification level that is set by the user—this will become clearer in just a moment.

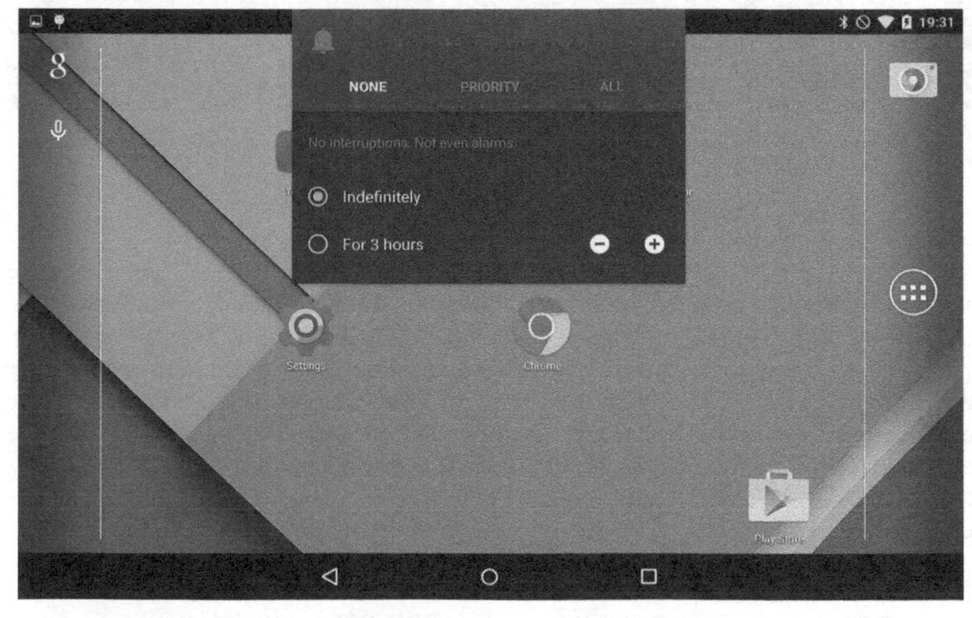

Figure 6-6A Sound and Notification, quick control widget—Priority level NONE.

Figure 6-6A shows the Notification filter set at NONE, which tells the Android OS that the user does not want any Notifications for the specified duration. You will notice that the volume slider has become invisible; this is because—with the Notification level set at NONE, the user will not receive any Notifications, alerts, phone calls, and calendar reminders—the volume level is rendered irrelevant.

The icon associated with the Notification level of NONE is a circle with a diagonal line across it, which is shown on the status bar indicator during the interval that this mode is set for. This indicator makes the user aware of the current Notification Priority setting.

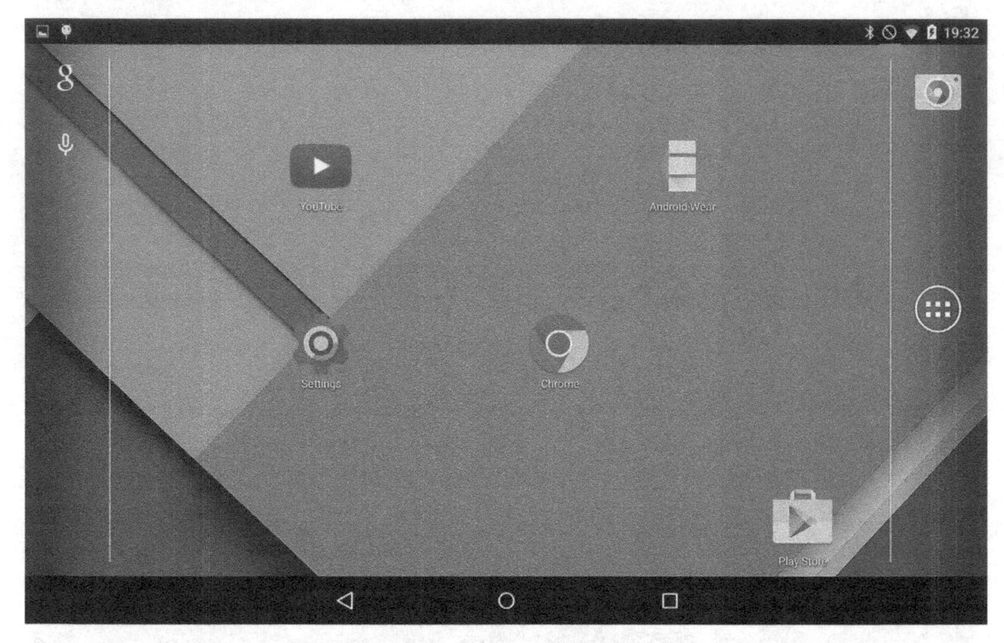

Figure 6-6B Sound and Notification, status indicator for Notification level NONE.

Figure 6-6B shows the icon for Notification level NONE in the status bar.

Probably the most useful Notification level is the PRIORITY setting. Users who desire not to be interrupted with routine Notifications can set the Notification level at PRIORITY along with the desired time duration. This setting tells the Android OS that only those Notifications set with the Priority flag in the originating App's application code (and/or applications and contacts whose Notifications have been marked as Priority in the *Settings*—as we will see in the next sections) will get through to the user during the specified interval while applying the volume level as set in the slider.

Figure 6-7A shows the widget setting at PRIORITY with the duration set at 4 hours—during which period, only the highest-priority Notifications will get through to the user.

Also, the five-pointed star—used as the visual indicator for Priority mode—is displayed in a Toast and also included in the status bar at the top right of the device.
Figure 6-7B shows the icon for Notification level PRIORITY in the status bar.

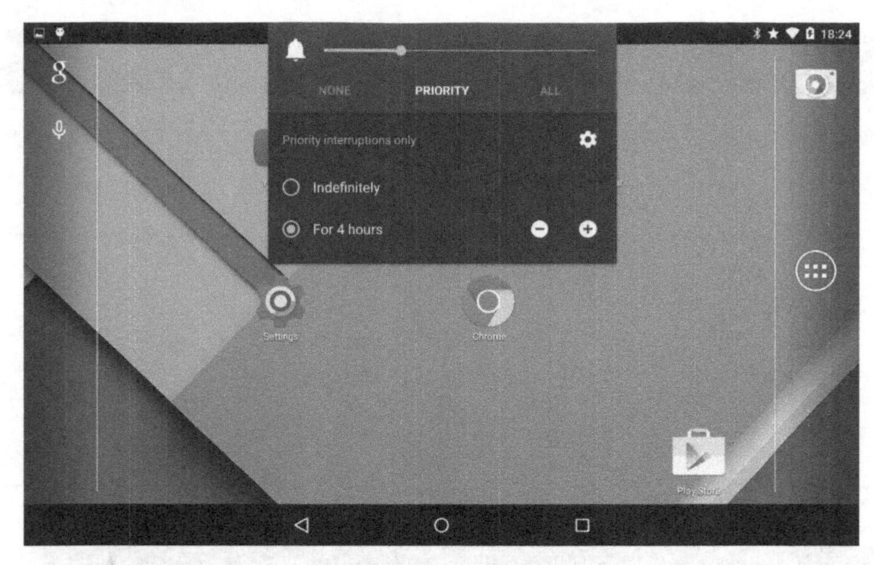

Figure 6-7A Sound and Notification, Notification level PRIORITY.

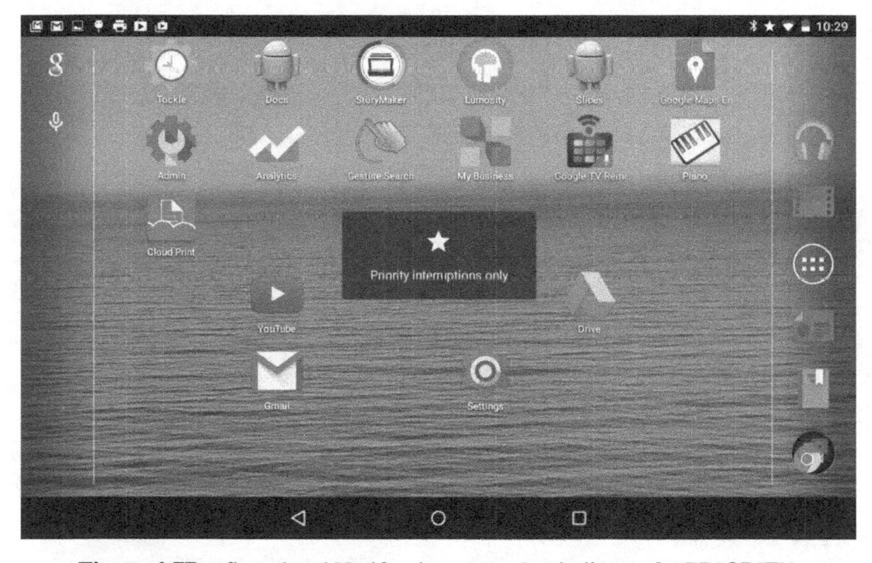

Figure 6-7B Sound and Notification, status bar indicator for PRIORITY.

6.4.2 Notification Configuration and Control

Besides this quick Notification control widget we have just covered, you will find that the OS/Device *Settings* have an item, *Sound and Notification*, that provides access to a wide range of configuration and control.

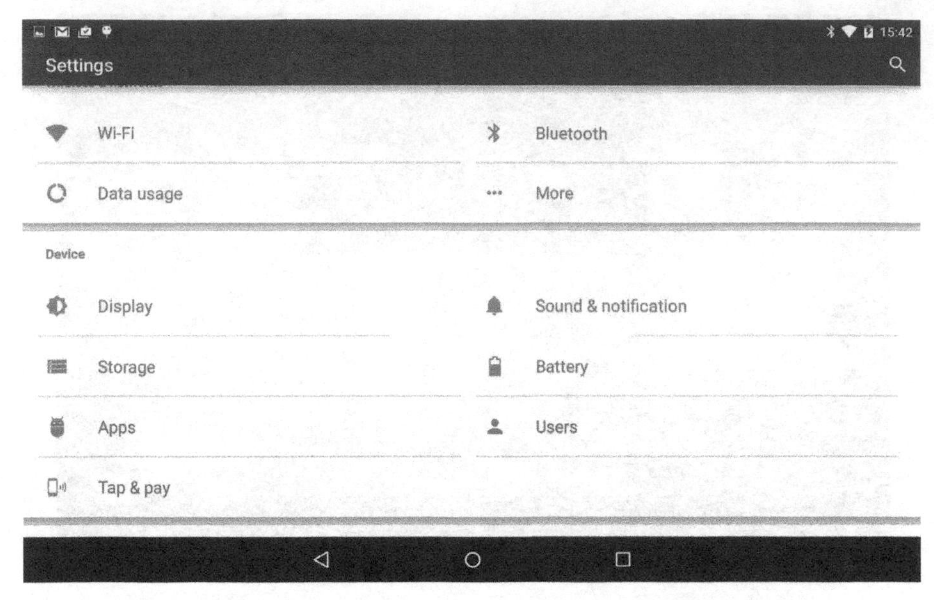

Figure 6-8 Sound and Notification in OS Settings.

Figure 6-8 shows both the Sound and Notification item in Device Settings.

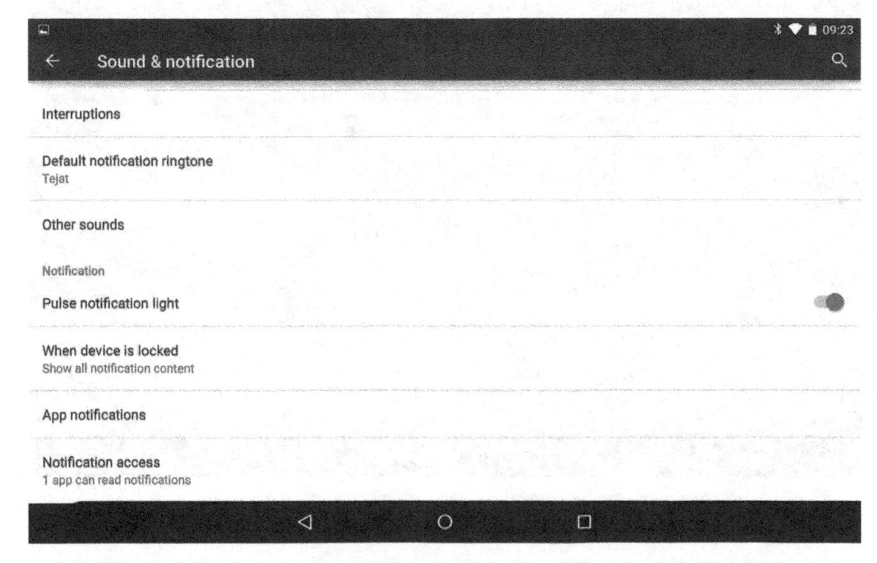

Figure 6-9 Notifications and Interruptions in Sound and Notification.

Within *Sound and Notifications*, there are several subitems of which *Notifications* and *Interruptions* are of special interest. Figure 6-9 shows the subitems under *Sound and Settings*. The boolean *Pulse Notification Light* setting enables the blinking of your device's LED when a Notification is received during a time when your devices is idle/sleeping. The blinking light helps indicate to the user that some new Notification has arrived such as calendar appointment, email, a missed call, etc.

6.4.3 Locked Screen and Notifications

In general, users can protect their privacy using the options such as a screen lock—available under *Screen Security* in the *Security Settings*. Related to the screen lock, the item *When Device is Locked* under *Notifications* helps users control the behavior of Notification delivery while the screen is locked.

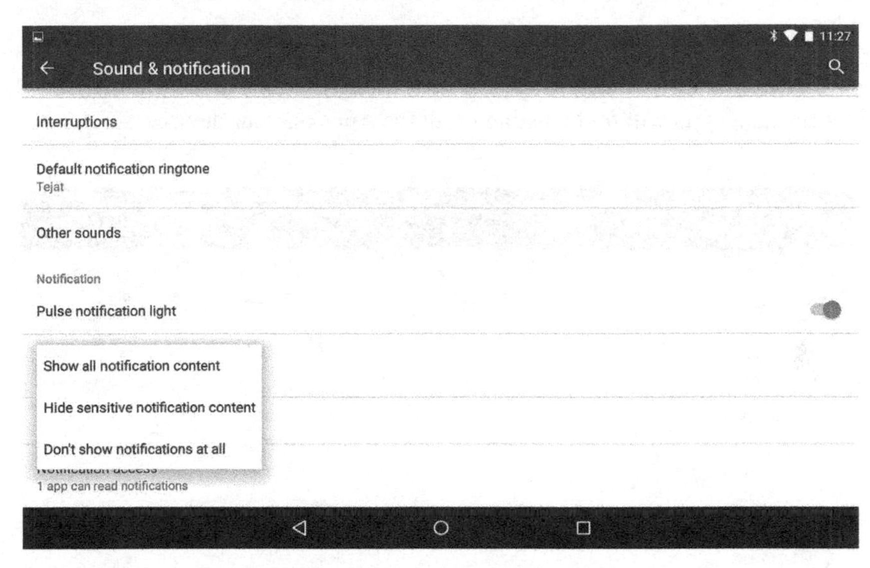

Figure 6-10 Notification privacy control options, when device is locked.

Figure 6-10 shows the lock screen privacy options—a user can choose either not to have any Notifications shown past the locked screen or to have them shown—with or without

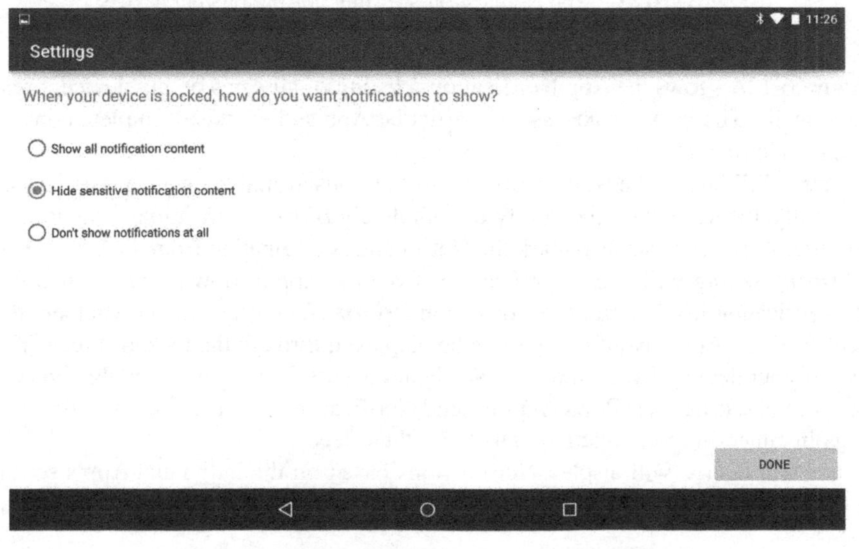

Figure 6-11 Postscreen lock setup, Notification options.

display of sensitive content. In case of a Notification about a missed call, the details of the caller and/or the contents of the voice message are examples of sensitive content.

Also, right after users set up their *Screen Lock* under *Security* → *Settings*, they are presented with Notification-related options.

Figure 6-11 shows the Notification options offered right after users set up their screen lock under security, thus giving users complete control of Notification behavior past the screen lock.

App Notifications is another sub-item under Notifications and of special interest to us, because it governs the behavior of any App's Notification on the user's Wear device. Under App Notifications, you will find a listing of all the Apps on your device.

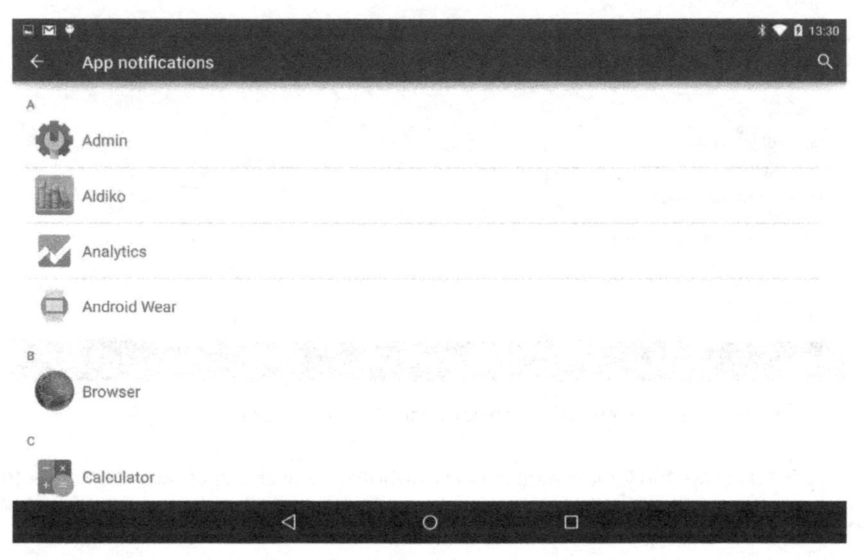

Figure 6-12A App Notifications—listing of Apps on your device.

Figure 6-12A shows the App Notifications' listing of all Apps on my device, organized alphabetically. The user can access any particular App and exercise complete control over that App's Notifications.

Figure 6-12B shows the Notification control options available on a per App basis. Most significantly, the users have the ability to completely *Block* Notifications originating from a particular App. They can also mark the Notifications originating from an App as *Priority*. The *Priority* setting will allow Notifications from the App to flow through when the user has a Notification level set for their device at *Priority*. The user can set whether sensitive content from the App's Notifications can be displayed through the locked screen. You will notice on your device that all the App Notification settings have a reasonable, good faith-based, neutral setting that allows Apps to send Notifications. Users can control these default settings in either direction relative to that, per their needs.

Your Wear device will display Notifications based on the individual App's setting. So even if your handheld device is set at Notification level NONE (effectively muted), your

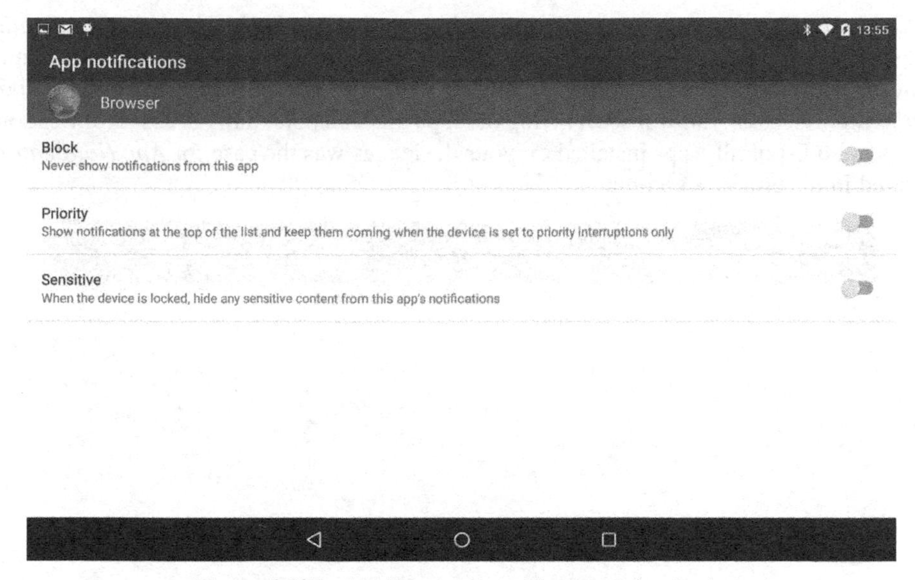

Figure 6-12B Notification control on a per App basis.

Wear device can still vibrate if it has not been muted. Effectively, your handheld device and your Wear device can be muted independently.

Also, while developing and debugging Wear Apps, you will need to ensure that you do not inadvertently *Block* the *Android Wear* App (Figure 6-12C).

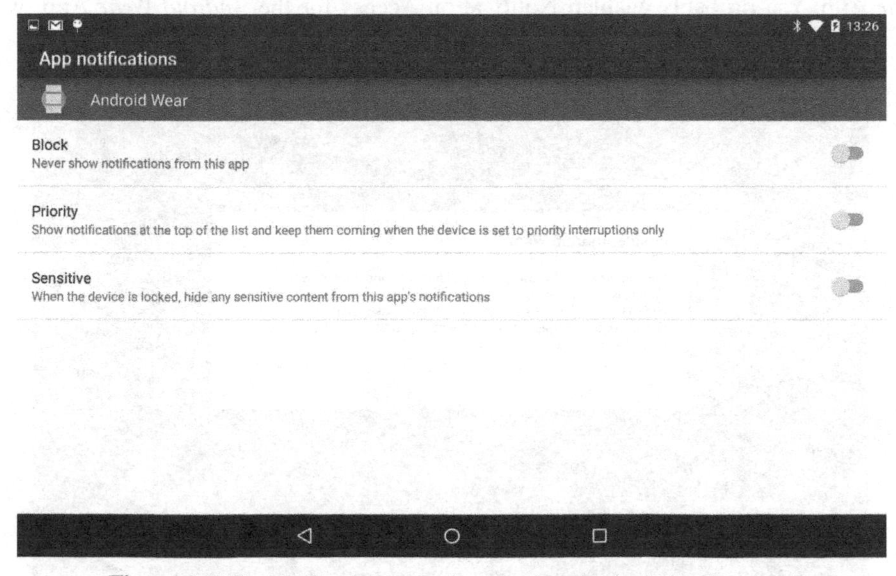

Figure 6-12C Notification control settings on the Android Wear App.

6.4.3.1 Notification Access Another subitem under *Notifications* is *Notification Access*, which is distinct from *App Notifications*. *Notification Access* helps the user control which Apps can access Notifications posted by the system or by other Apps. Under *Notification Access*, you will see a listing of Apps that can potentially access Notifications; this is not a list of all Apps installed on your device, as was the case for *App Notifications* covered in the previous section.

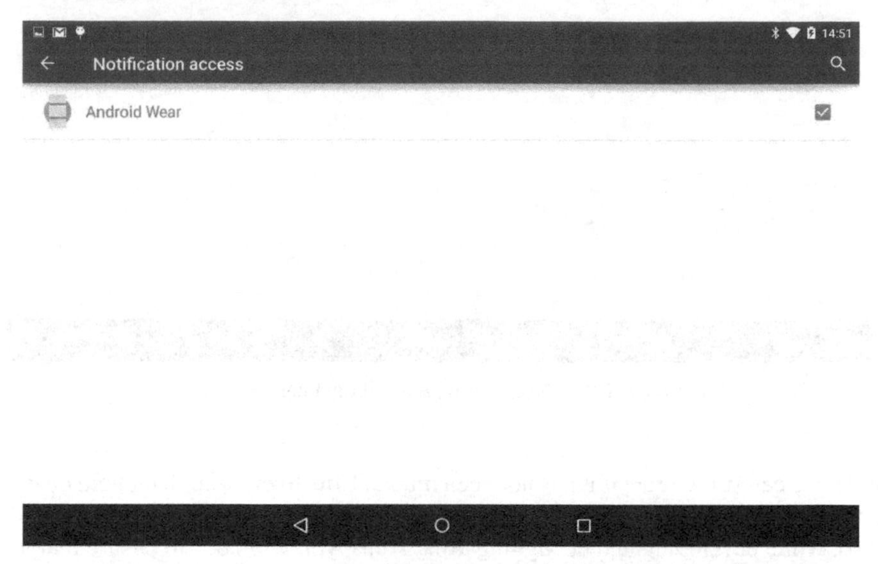

Figure 6-13A Notification Access—list of Apps with enabled/disabled status.

Figure 6-13A shows the list of such Apps on my device, which happens to be the *Android Wear* App. You probably enabled Notification Access for the *Android Wear* App, when

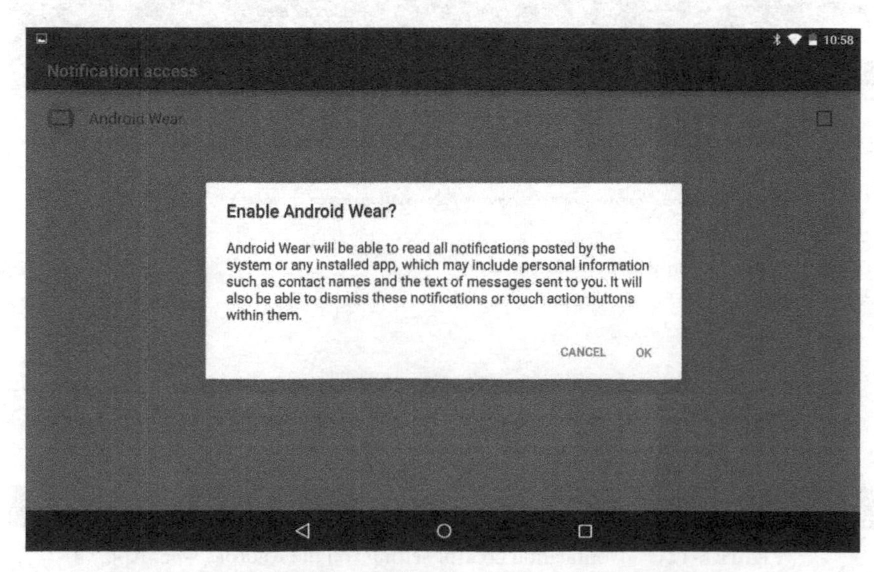

Figure 6-13B Notification Access—list of Apps.

running it for the first time. If you toggle this setting, you will see the kind of notice dialog that pops up. This is obviously a very powerful setting because the App can access all Notifications—from all other Apps as well as the Android OS. Users will do well to grant this access only to Apps that they trust.

Figure 6-13B shows the dialog that reveals the intricacies of allowing Notification Access for any App. Your App too can access Notification info if it has a good reason to, by implementing the *NotificationListenerService*, which resides in the *android.service.notification* package.

6.4.4 Interruptions

Another item of relevance to us as developers is the *Interruptions* item under *Sound and Notification. Interruptions* is another layer of control that users can exercise over the intrusive aspect that is inherent to Notifications. Users can indicate their downtime days and times—during which only Priority Interruptions will reach them.

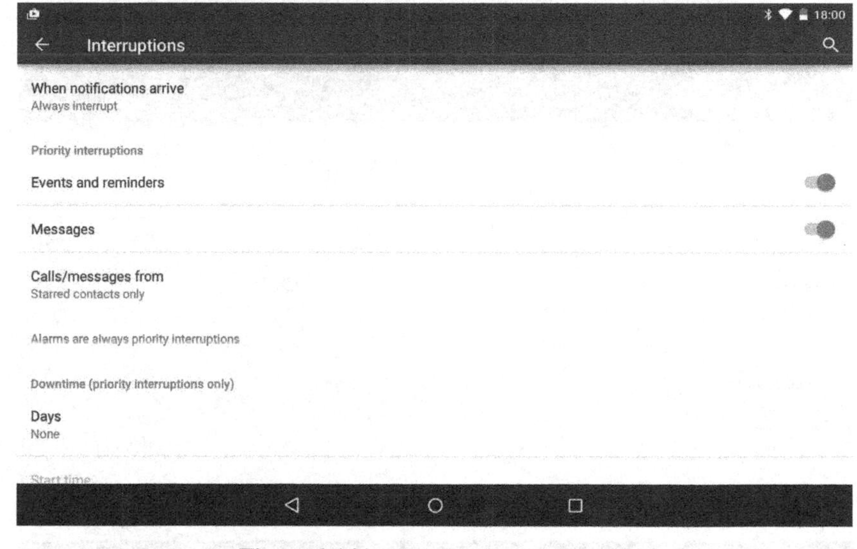

Figure 6-14A Interruptions, subitems.

Figure 6-14A shows the various subitems under Interruptions.

Figure 6-14B shows the options under the item *When a Notification arrives*, namely, *Always Interrupt*, *Allow Only Priority Interruptions*, and *Don't Interrupt*.

The user also has the ability to toggle whether *Events and reminders* and *Messages* are to be included in *Priority Interruptions*.

Figure 6-14C shows the options under *Calls/Messages from*, namely, *Anyone*, *Starred contacts only*, and *Contacts only*.

Figure 6-14D shows the item *Downtime days*, which helps users pick their downtime days as well as the start time and end time.

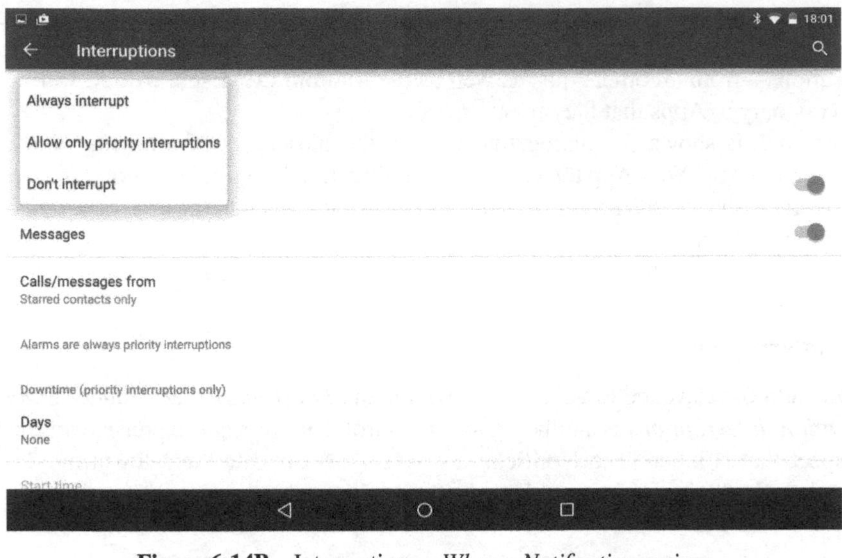

Figure 6-14B Interruptions—*When a Notification arrives.*

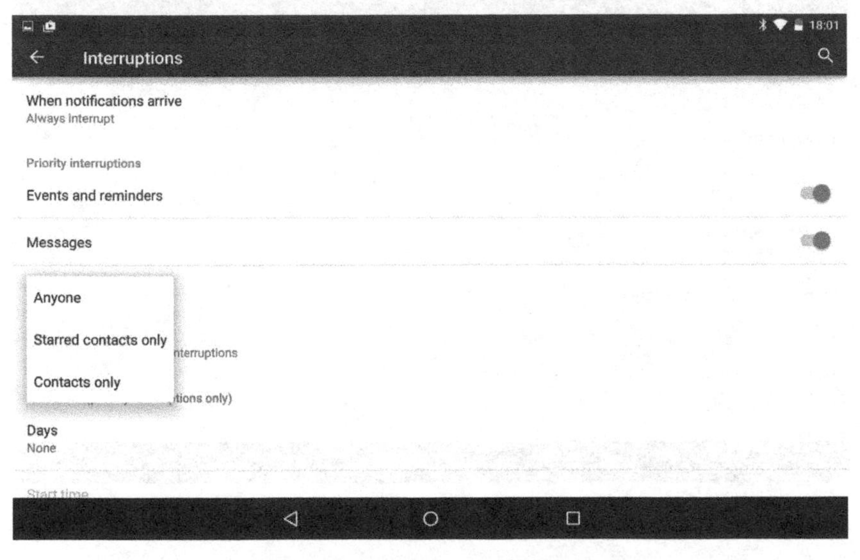

Figure 6-14C Interruptions—*Calls/Messages from.*

6.5 App Notification Strategy

It behooves any App developer to avoid the casual and frequent use of Notifications. If an App comes across as being too "aggressive" or "noisy" with regard to Notifications, the user may find it bothersome and therefore decide to avail of the ability to turn Notifications off for that App or even uninstall the App altogether. The best strategy with regard to

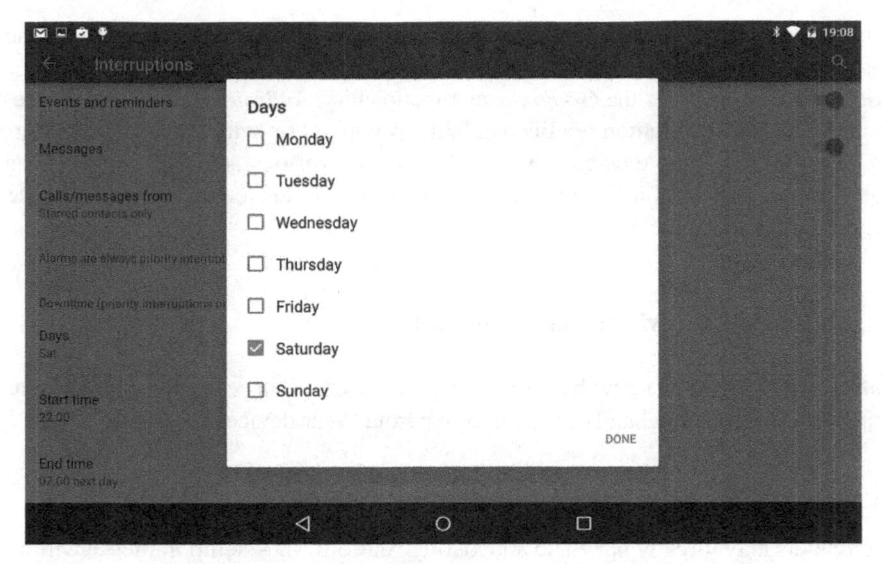

Figure 6-14D Interruptions—*Downtime days.*

Notifications is for Apps to use a conservative approach, possibly coupled with an adaptive and gradual, "back off" or "accelerate" approach based on analyzing the usage metrics. If a user has been dismissing an App's Notification without consuming it, by entering the App, that's a sign that the Notification was likely not interesting enough or welcome. Using such a data- and evidence-based approach can personalize and align the Notification threshold at the level of positive user engagement. Apps can also present users with the option to turn Notifications off from within the App so that they do not originate them in the first place if the user is not interested in them. Oftentimes, it's a matter not of having Notifications on or off, but about finding the right quality, level, and category of Notifications that engages the user positively.

6.6 Google Now and Android Wear

Google Now is an intelligent personal agent or assistant from Google that works in the background to bring you relevant contextual information, where and when you are most likely to need it. Google Now emphasizes a voice-based and hands-free experience. *Google Now* organizes information in the form of simple Info Cards on weather, next appointments, commute times, parking places, traffic information, boarding passes, stocks quotes, sports scores, and more.

Google Now emphasizes the Suggest paradigm by making timely and contextually relevant suggestions to the user while also making the Demand-based voice search and user input lightweight and convenient. There are strong overlaps of purpose between the *Android Wear* platform and *Google Now*.

Based on the context of where you are, such as an airport, and based on the information that is available with Google, such as an airline ticket in your recent email, *Google Now* provides you a Notification/suggestion card with your boarding pass at the right time and place—without you having to search for it in your email.

Android Wear includes the *Google Now* functionality. *Android Wear* aims to make contextually relevant information readily available as you get on with your day, move around from place to place, and engage in your schedule and various activities. Information can become relevant based on where you are and what you are engaged in, and these can change moment to moment.

6.7 Android Wear Devices: Getting Started

It's always exciting to get new hardware setup, connected and ready for development. We will now engage in some hands-on setup of Android Wear devices for development.

6.7.1 Android SDK Wear Platform updates

Some readers may already have had an existing Android SDK setup in their environment, and if so, it is important to ensure that both the *Android Wear* platform (API level 20) and the Android 5.0/Lollipop (API level 21) are available in your local Android SDK environment. *Android Wear* requires Android 4.3 and above; however, this book targets the Android 5.0/Lollipop (API level 21) for all the examples and sample code (Figure 6-15).

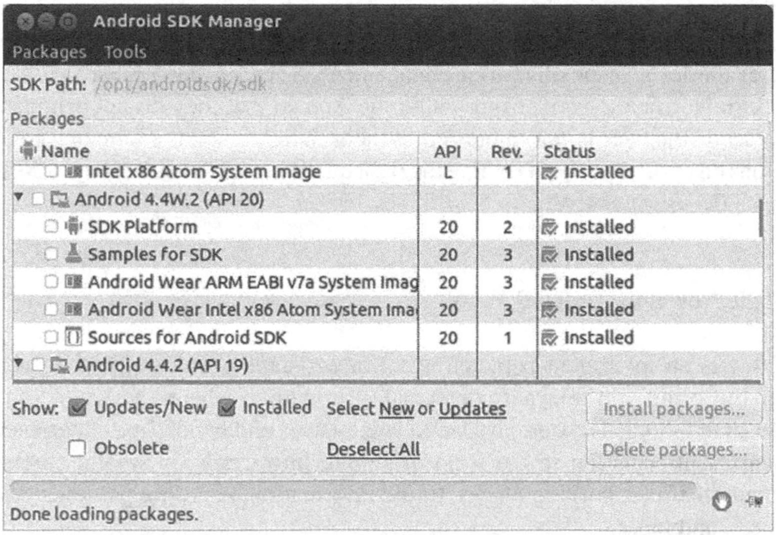

Figure 6-15 Android SDK Manager, Wear platform API level 20.

Running the command `android list targets` is another way to verify that you have the *Android Wear* platform in your Android SDK environment. The output of the command should show content that includes the *Android Wear* platform/API level, as shown in the snippet below:

```
id: 7 or "android-20"
    Name: Android 4.4W.2
    Type: Platform
    API level: 20
    Revision: 2
    Skins: HVGA, QVGA, WQVGA400, WQVGA432, WSVGA, WVGA800
(default), WVGA854, WXGA720, WXGA800, WXGA800-7in,
AndroidWearRound, AndroidWearSquare, AndroidWearRound,
AndroidWearSquare
Tag/ABIs : android-wear/armeabi-v7a, android-wear/x86
```

6.7.2 Procuring an Android Wear device

There are several *Android Wear* device models available on the Google Play Store: https://play.google.com. Figure 6-16 shows several *Android Wear* devices from ASUS, LG, Sony, Motorola, and Samsung. The prices, at the time of writing, start at the $199 price point. Over time, it is likely that the prices will reduce. The product reviews and ratings on the Play Store may also be useful for choosing a device to purchase.

Figure 6-16 Android Wear devices available on Google Play Store.

The following sections cover the steps involved in readying your *Android Wear* device for software development.

6.7.2.1 Using Android Emulator with Wear AVD Running the *Android Wear* Android Virtual Device (AVD) in the Emulator can be useful in case you do not have a Wear device.

One way to start the *Android SDK Manager* is by running the `android` command from your command line. Once the *Android SDK Manager* is started, you can access the *Tools* menu under which you will find the Manage AVDs item. Clicking on the Manage AVD will take you to the *Android AVD Manager*.

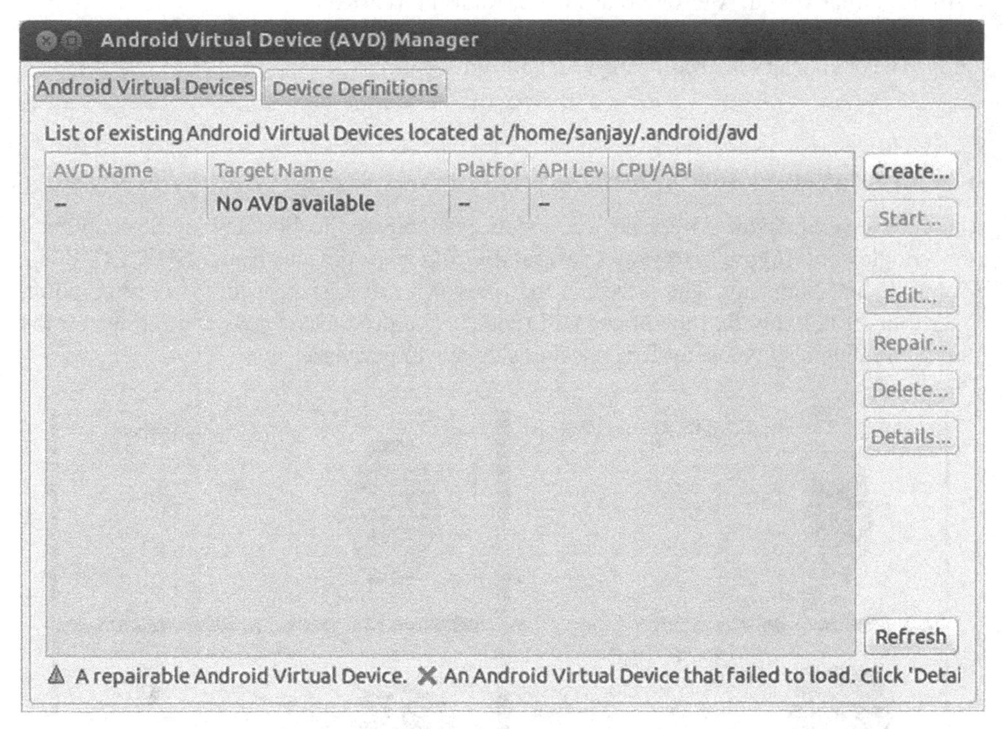

Figure 6-17A Android Virtual Device (AVD) Manager.

Figure 6-17A shows the AVD Manager screen. Clicking on the *Create* button on the right will take you to the *Create New AVD* screen.

Figure 6-17B shows the Create New AVD screen. You can choose a name of your choice. You will need to select various options as indicated in the figure. The target will need to be a current Wear platform—*Android Wear 4.4w.2* (API level 20) at the time of writing. After filling in all the fields, clicking on next will get you to the "result" screen.

Figure 6-17C shows the output result summary info of the AVD creation step. Clicking on OK will take you back to the AVD Manager main screen. You should see your recently created Wear AVD.

Figure 6-17D shows the recently created AVD listed. You can select it and click on the *Start* button toward the right.

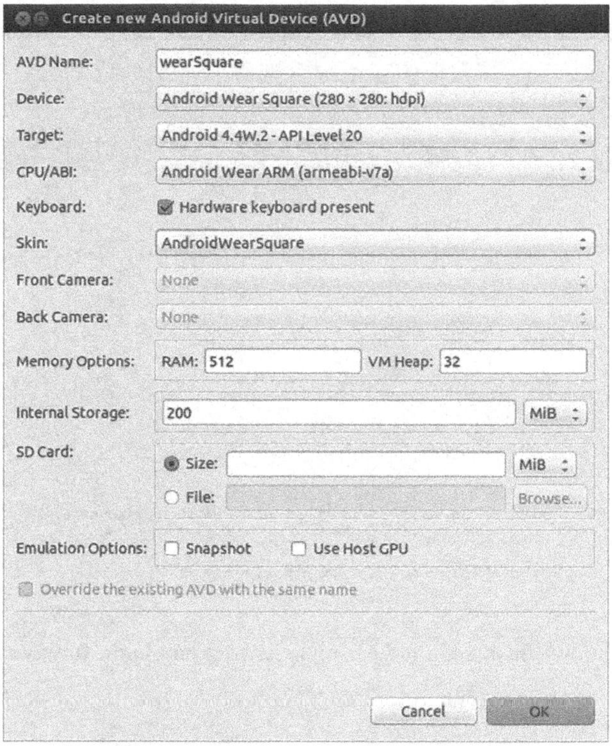

Create new Android Virtual Device (AVD)
AVD Name: wearSquare
Device: Android Wear Square (280 × 280: hdpi)
Target: Android 4.4W.2 - API Level 20
CPU/ABI: Android Wear ARM (armeabi-v7a)
Keyboard: Hardware keyboard present
Skin: AndroidWearSquare
Front Camera: None
Back Camera: None
Memory Options: RAM: 512 VM Heap: 32
Internal Storage: 200 MiB
SD Card:
Size: MiB
File: Browse...
Emulation Options: Snapshot Use Host GPU
Override the existing AVD with the same name
Cancel
OK

Figure 6-17B Create a new Wear AVD.

Android Virtual Devices Manager
Result of creating AVD 'wearSquare':
Created AVD 'wearSquare' based on Android 4.4W.2, Android Wear ARM (armeabi-v7a) processor,
with the following hardware config:
disk.dataPartition.size=200M
hw.accelerometer=yes
hw.audioInput=yes
hw.battery=yes
hw.dPad=yes
hw.device.hash2=MD5:f0b5528a2769367d5624a75b0c789fe4
hw.device.manufacturer=Google
hw.device.name=wear_square
hw.gps=yes
hw.keyboard=yes
hw.keyboard.lid=yes
hw.lcd.density=240
hw.mainKeys=yes
hw.ramSize=512
hw.sdCard=no
hw.sensors.orientation=yes
hw.sensors.proximity=yes
hw.trackBall=no
skin.dynamic=no
vm.heapSize=32
OK

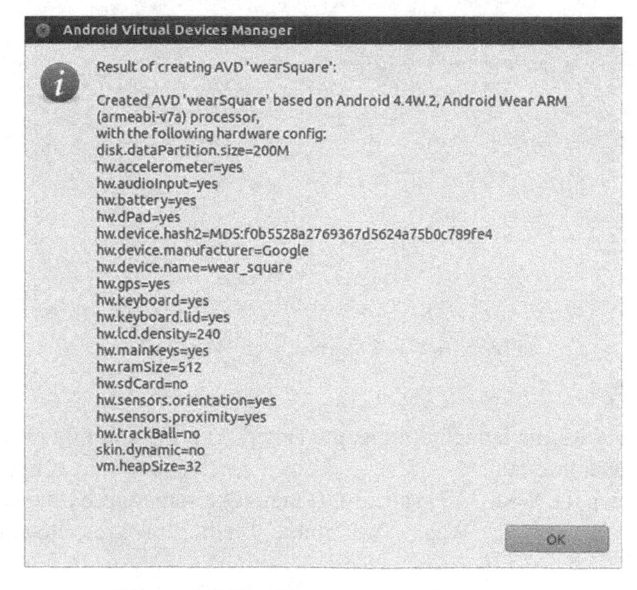

Figure 6-17C Wear AVD creation result.

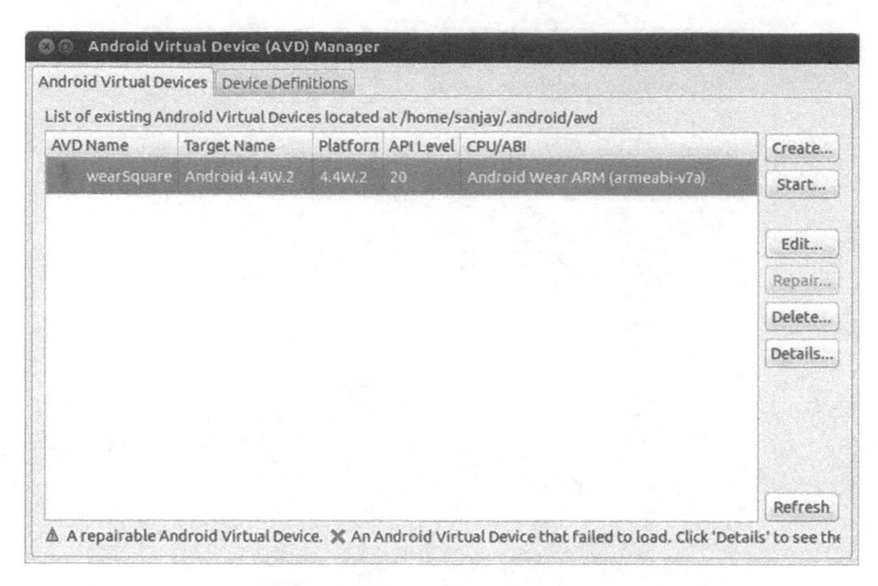

Figure 6-17D Start your Wear AVD.

Figure 6-17E shows the screen indicating a starting emulator. It may take some time to progress.

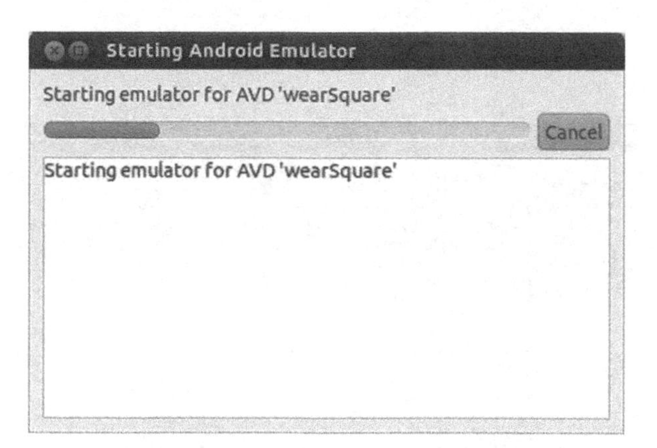

Figure 6-17E Starting your Wear AVD.

Figure 6-17F shows the launch options presented. You may simply go with the default and click the *Launch* button.

Figure 6-17G shows the Wear AVD starting. It may take some more time to progress.

Figure 6-17H shows the Wear AVD make further progress toward completing startup.

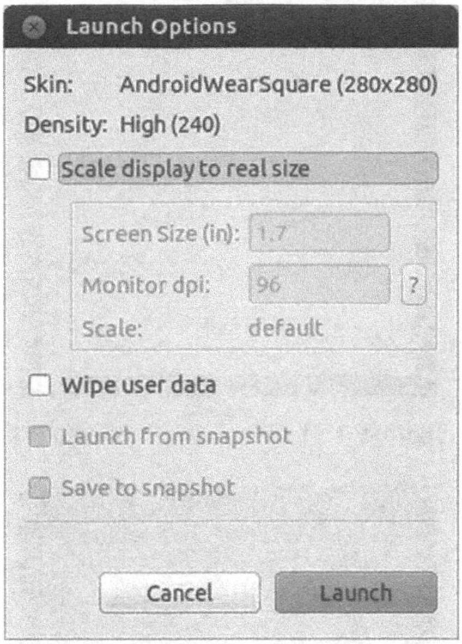

Launch Options

Skin: AndroidWearSquare (280x280)

Density: High (240)

Scale display to real size

Screen Size (in): 1.7

Monitor dpi: 96

Scale: default

Wipe user data

Launch from snapshot

Save to snapshot

Cancel Launch

Figure 6-17F Wear AVD launch options.

Figure 6-17G Wear AVD screen during boot.

Figure 6-17I shows the successfully started Wear AVD.

Figure 6-17J shows the step of pairing your handheld device with your running Android Wear AVD.

Figure 6-17H Wear AVD nearly started.

Figure 6-17I Wear AVD running.

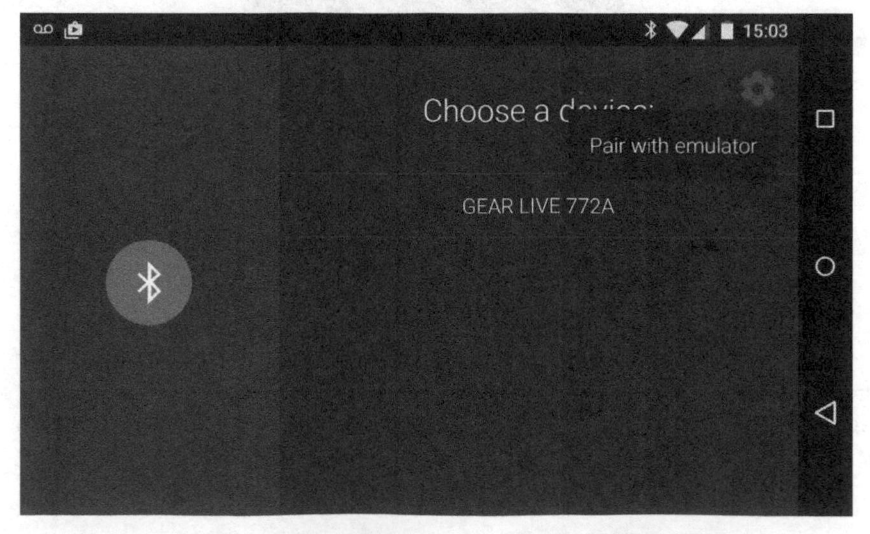

Figure 6-17J Pairing with running emulator.

6.7.3 Pairing and Enabling Developer Mode

Any new Wear device will typically need to be paired with a handheld device. As developers, we will need to also get our Wear device set up for development and debugging. This section covers the steps for pairing and enabling developer mode from scratch.

6.7.3.1 Unboxing your Wear device In case you have already paired and enabled Developer mode on you Wear device, you may skip this section:

1. **Unbox** your *Android Wear* device and perform the initial charging and/or other steps per the manufacturer's instructions.
2. **Power up** your *Android Wear* device for the first time, per the manufacturer's instruction set. After you boot your Wear device for the first time, you will be presented with a home screen.

Figure 6-18 Android Wear device watch face, postboot.

Figure 6-18 shows a home screen watch face with status indicator icons for Notification Sync status, Pairing status, and the battery level. You will likely observe that if your watch is idle, the screen dims. Wear devices enter a low-power, ambient mode when not being used—in order to conserve battery power. Tapping anywhere on a dim screen will wake up your watch.

6.7.3.2 Pairing your Handheld device with your Wear device Once the *Android Wear* device boots successfully, you will need to **pair** the *Android Wear* device with your handheld Android phone or tablet device. This can be accomplished by installing and running Google's *Android Wear* App, which is available on the Google Play Store, on your handheld mobile Android phone or tablet device. A search for keyword "Wear" on the Google Play App Store should get you results that include the App named *Android Wear* by Google, Inc. The link to the *Android Wear* app is https://play.google.com/store/apps/details?id=com.google.android.wearable.app.

Once you have installed the *Android Wear* App, you will need to run it while keeping both your handheld device and mobile device close together. Bluetooth will need to be turned on in the handheld device via the *OS Settings* in order to complete the detection and pairing steps.

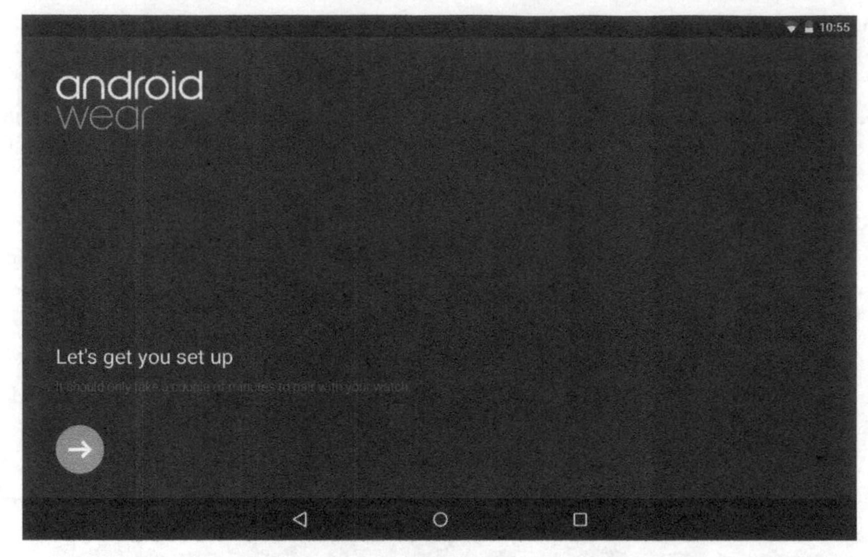

Figure 6-19A Android Wear App, running on a handheld Android device.

Figure 6-19A shows the *Android Wear* App on its first run.

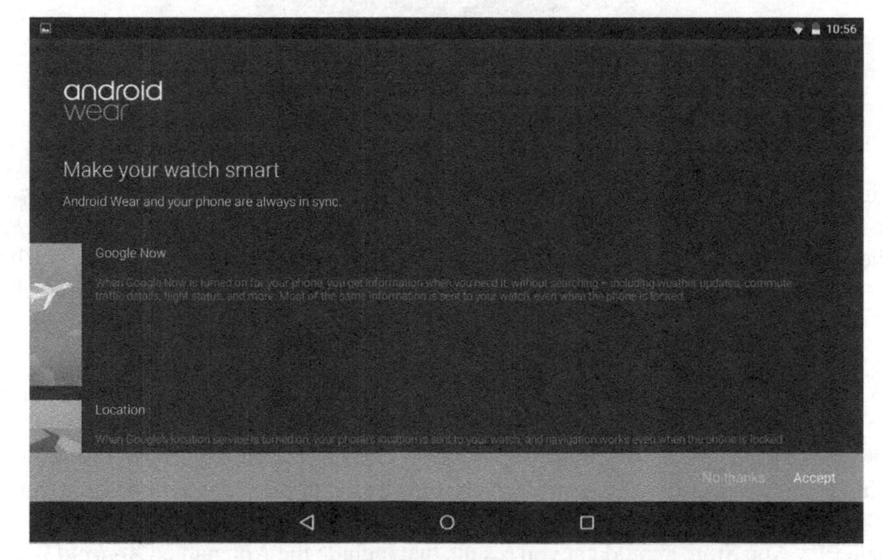

Figure 6-19B Android Wear App, introductory screens.

Several introductory screens will guide you through the steps of detecting your Wear device and pairing with it from your handheld device (Figure 6-19B).

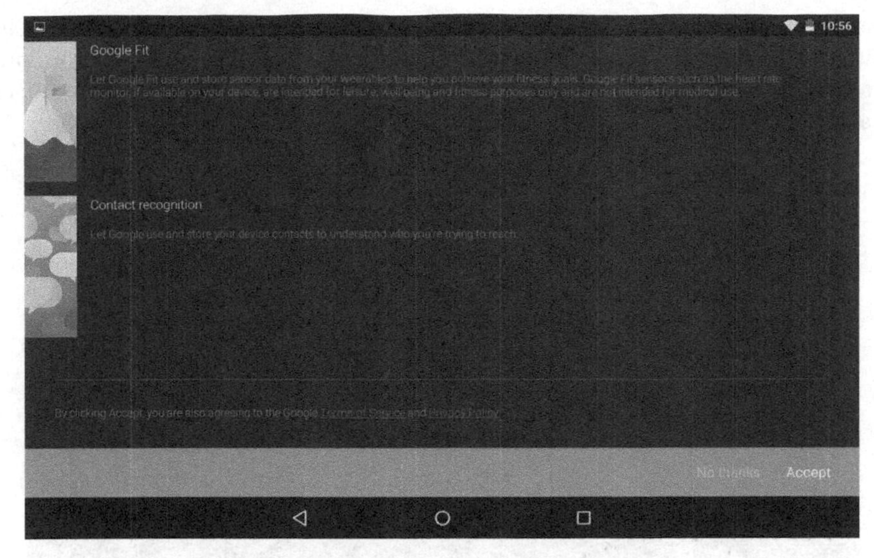

Figure 6-19C Android Wear App, several introductory screens.

You will notice that *Google Now* and *Google Fit* can be integrated with the *Android Wear* experience (Figure 6-19C).

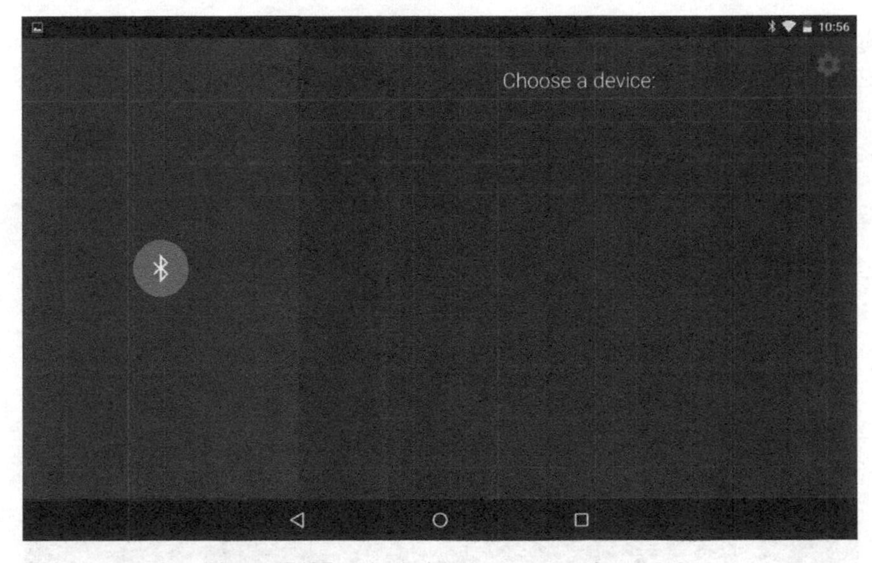

Figure 6-19D Android Wear App, device scan.

It may take a moment for your handheld device to detect your Wear device over Bluetooth (Figure 6-19D).

Once your Wear device is detected and listed, as shown in Figure 6-19E, you can tap on the Wear device model (such as *Gear Live 772A* seen in the figure) in order to initiate connectivity. You should also keep an eye on your Wear device screen for any changes or display of information.

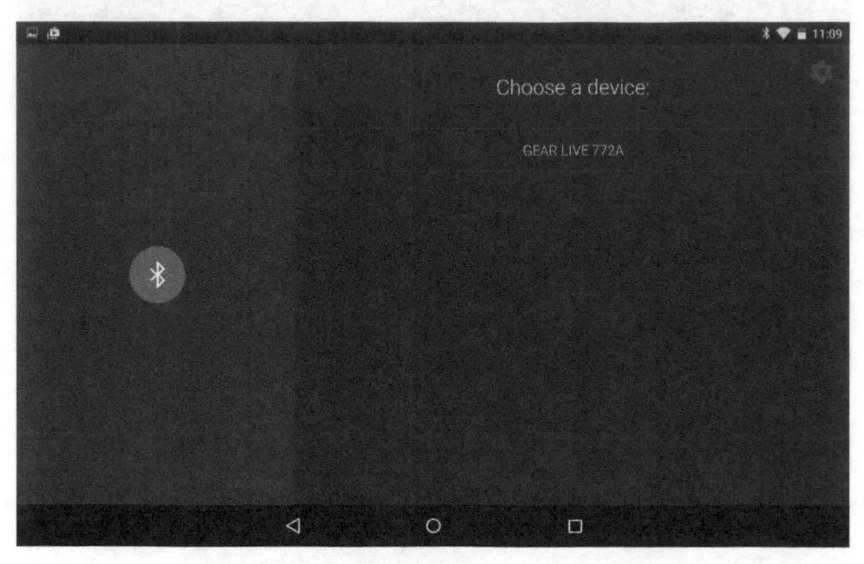

Figure 6-19E Android Wear App, device scan contd.

Figures 6-19F and 6-19G show the progressive screens that you will see during this process.

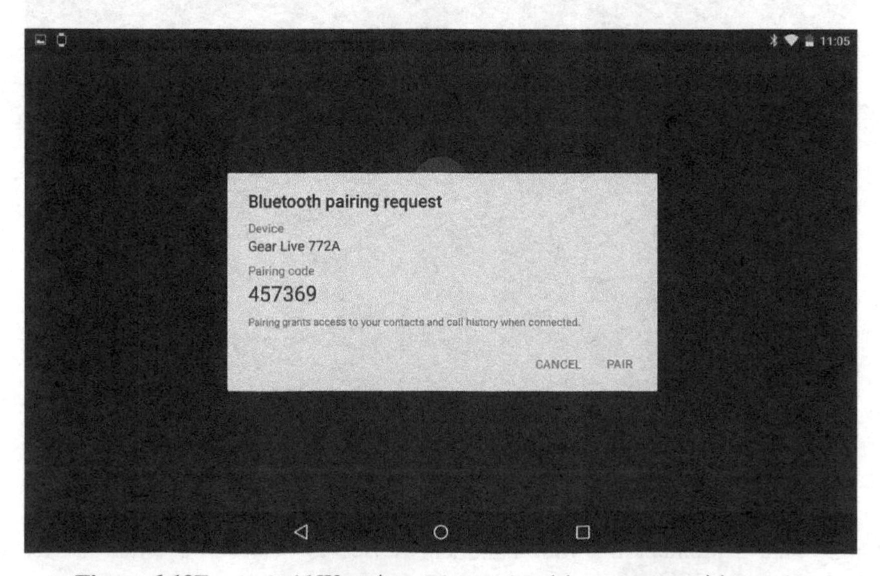

Figure 6-19F Android Wear App, Bluetooth pairing request, pairing code.

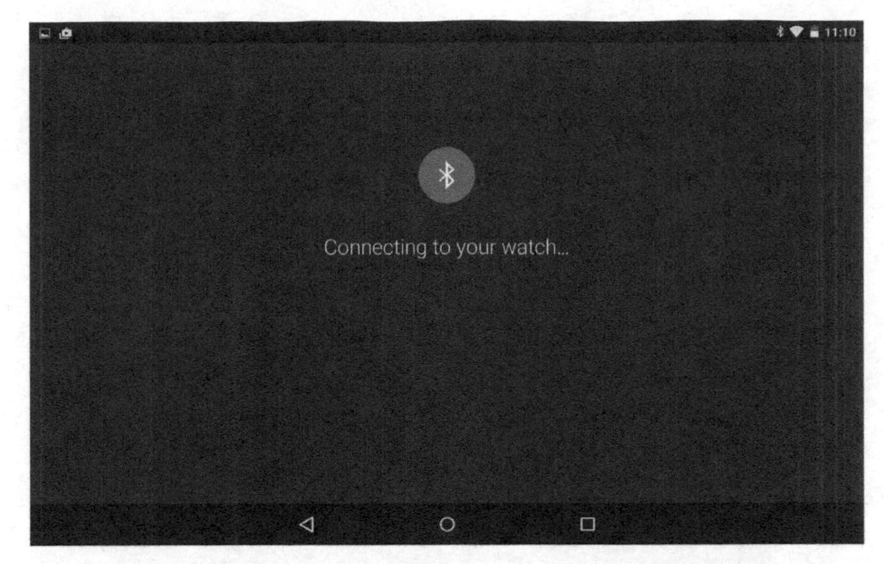

Figure 6-19G Android Wear App, connecting to your Wear device.

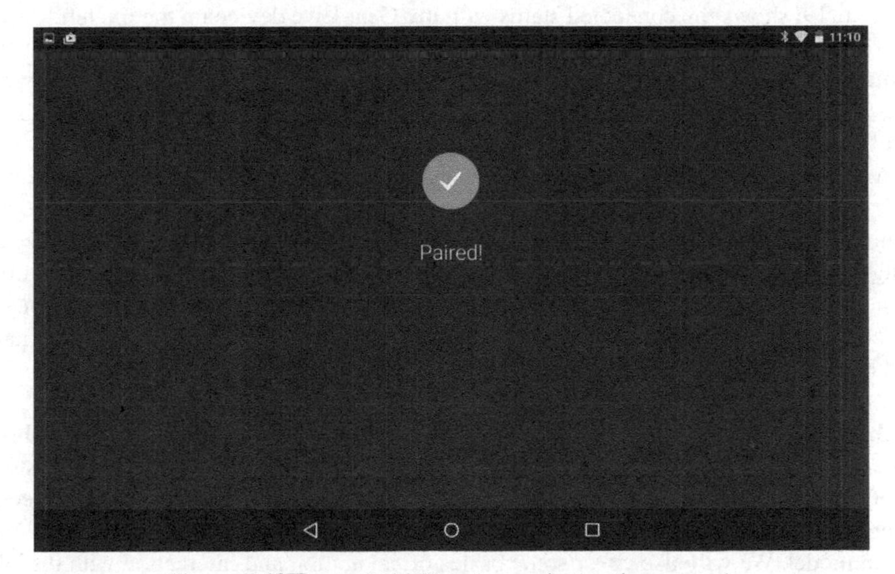

Figure 6-19H Android Wear App, paired and connected.

Figure 6-19H shows the screen that indicates successful pairing.

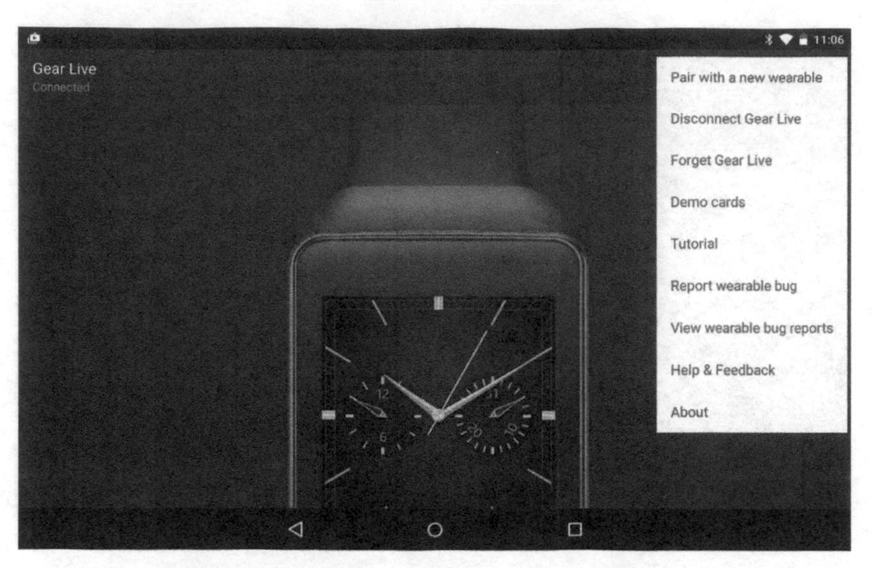

Figure 6-19I Android Wear App, successful pairing.

Figure 6-19I shows the connected status with my Gear Live device on the top left.

You will also find a gears icon on the top right of the action bar as well as the overflow menu item on the action bar. The options within the overflow menu are shown in Figure 6-19I. There are tutorials and demos available here, as well as the ability to disconnect from the connected Wear device and so on.

You can only pair your watch with one handheld device at a time. If you have one Wear device and multiple phones, you can pair your Wear device with a different handheld by running the Wear App on your handheld, disconnecting from any current Wear device and using the option: *Pair to a new wearable*. If you have one Wear device and multiple phones, you may need to reset your Wear device to the factory settings first based on your device's instruction manual, after which you will be able to pair it with a different handheld device.

6.7.3.3 Enabling Developer Mode and Debugging Settings on your Wear device

Once you have paired the handheld device with your Wear device, the next useful step is enabling the *Developer mode* and *Debugging Settings* on your Wear device; this section covers the steps to accomplish this. Some specifics may vary depending on your Wear device model. We will also cover some basics of navigation and interaction with the Wear device that we will be needing to navigate the user interface in order to get debugging setup.

You will likely notice that the indicator denoting lack of tethered connectivity (cloud icon, with a line across it) disappears immediately after the successful pairing of the Wear device with your handheld device. Tapping on the home screen tells your *Android Wear* device that the user wishes to "Demand" something and takes you to the **Cue** Card, which is very appropriately named—the users are cuing or indicating to the device the specifics of their demand; the Wear device is seeking a cue or indication about the users' specific demand or needs.

Figure 6-20A Cue Card, initial display of voice interface.

Figure 6-20A shows the Cue Card with the voice interface, at the bottom of which lies an expansion icon—an upward pointing arrow. In the absence of any voice input for about 5 seconds, the Cue Card displays the options in this expanded list. You can also touch the expansion arrow icon to get to the expanded list of actions.

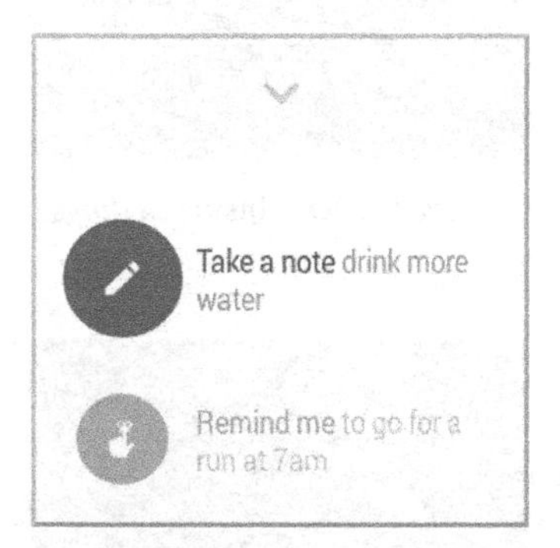

Figure 6-20B Cue Card, with various options in expanded list.

Figure 6-20B shows the expanded list of actions; you will notice that swiping downward will display more actions. The *Settings* action, which is of great interest for our needs, lies toward the bottom of this list of actions. Typically, you do not expect users to visit *Settings* very often. Part of good interaction design is to keep actions that are expected to be used most often more easily accessible, and vice versa.

Figure 6-20C shows the *Settings* action item. Tapping on *Settings* will take you to a list of subitems within Settings.

Figure 6-20D shows a few of the Settings' sublist of options. As you swipe upward and traverse down the list, you will see the various Settings items such as *Adjust Brightness*, *Bluetooth Devices*, *Always-On Screen*, *Airplane Mode*, *Restart*, *Reset Device*, *Change Watch face*, and *About*. The exact ordering and even some of the items may vary slightly based on your Wear device model.

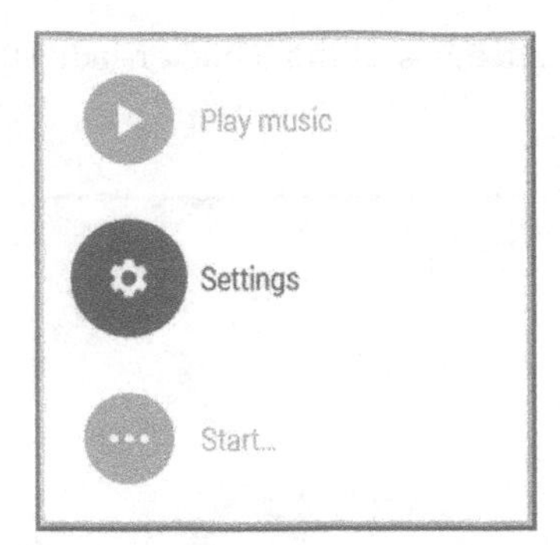

Figure 6-20C Accessing *Settings*.

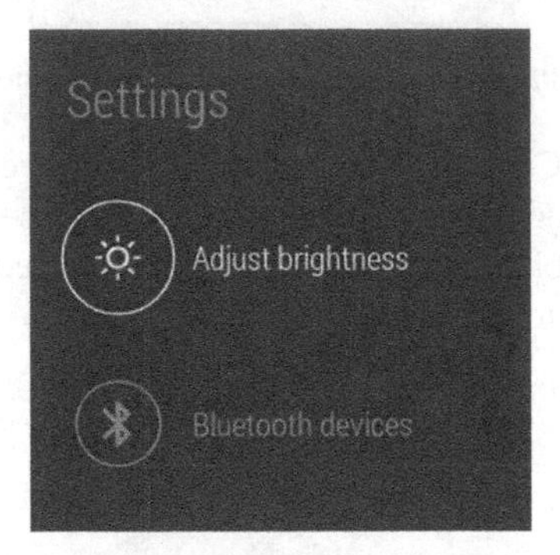

Figure 6-20D Inside *Settings*.

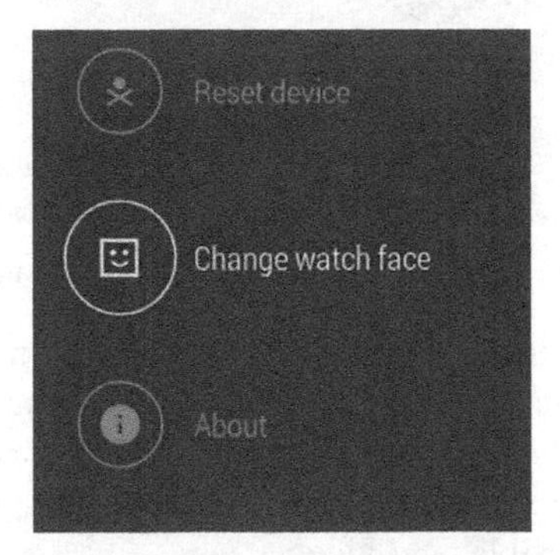

Figure 6-20E *Settings*, *About* item.

Figure 6-20E shows the *About* item at the bottom and one more upward swipe will select and highlight the *About* item. Touching the *About* item will take you into the *About* subitems.

Figure 6-20F shows the subitems under *About* within *Settings*, such as *Model*, *Software Version*, *Serial number*, *Build Number*, and so on. They represent system-level information

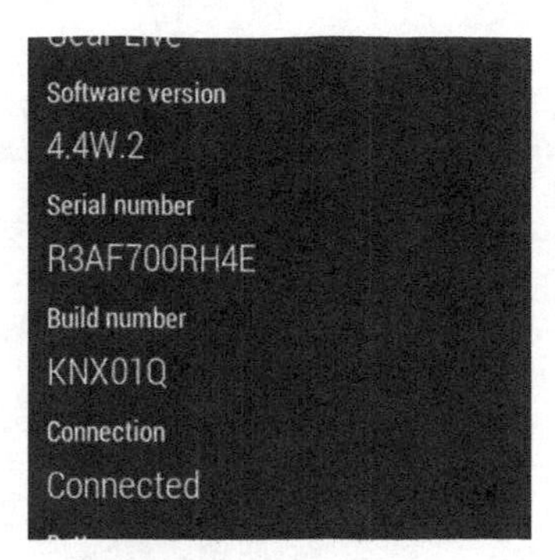

Figure 6-20F *Build number* in *About Settings.*

about the device and are of interest to us as developers. Tapping on the *Build Number* seven times will enable developer mode on your Wear device.

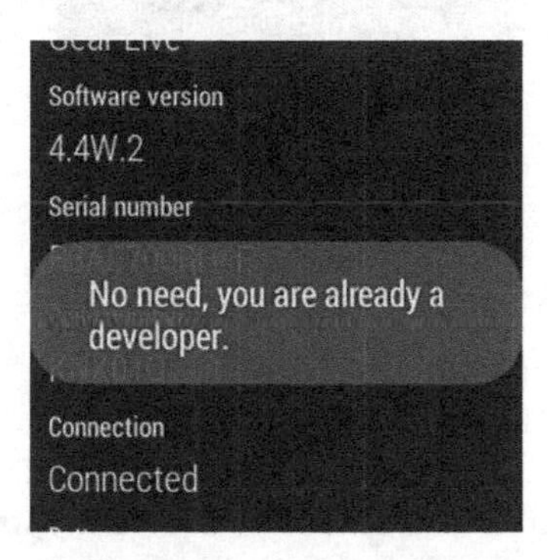

Figure 6-20G Tapping on the *Build number* in *About Settings.*

Figure 6-20G shows the effect of commencing tapping on the *Build Number* item on my device (for which Developer mode already happens to already be enabled). If you are doing this on a new device, you should see an appropriate message indicating that developer mode just got enabled based on your tapping seven times on the *Build Number.* As it turns out, enabling *Developer Mode/options* by tapping on the *Build Number* is a onetime step on a given device. Thereafter, you will have *Developer options* showing up on your device under *Settings*; you can always disable or enable it anytime by using a toggle switch within *Developer options.*

You must have observed by now that swiping to the right clears the current screen from the stack and gets you back to the previous screen. You can exit the *About* screen by swiping to the right.

Figure 6-20H shows the items under *Settings*, which now include the newly added *Developer options* item—after enabling Developer mode.

6.7.3.4 Enabling Wear ADB Debugging and Debug over Bluetooth Accessing the *Developer options* item will provide you with the options for enabling *ADB debugging* and *Debug over Bluetooth.*

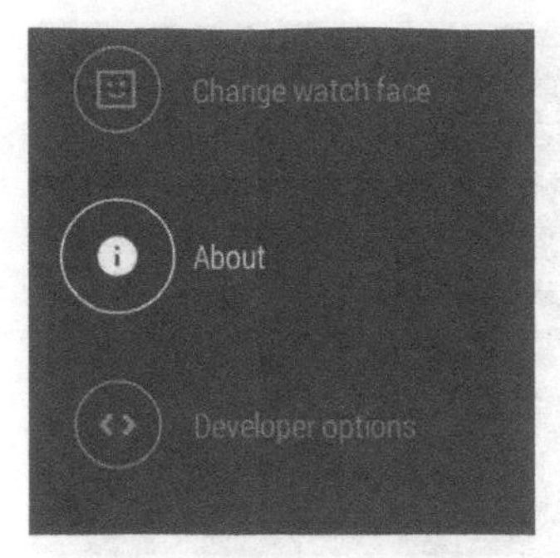

Figure 6-20H *Developer options*—item was added after enabling Developer mode.

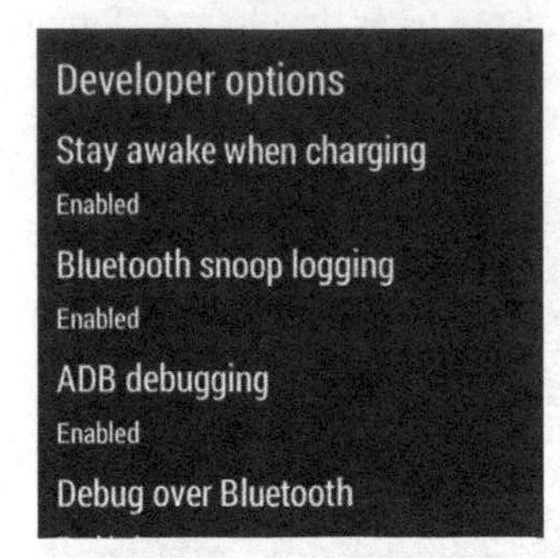

Figure 6-20I *Developer options.*

Figure 6-20I shows the various items under Developer options. The most significant of these are the *ADB Debugging* and *Debug over Bluetooth* items, both of which need to be enabled for development/debugging purposes.

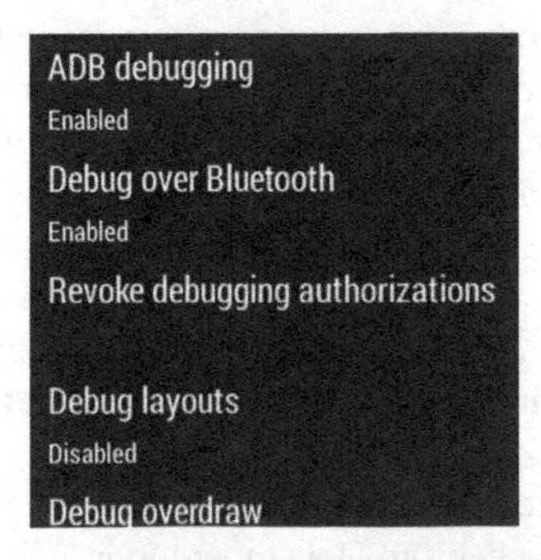

Figure 6-20J *Developer options*—enabling *ADB Debugging* and *Debugging over Bluetooth.*

Figure 6-20J shows that the items *ADB Debugging* and *Debug over Bluetooth* have been enabled.

6.8 Wear Debugging and Android SDK

Now that you have enabled Developer mode, ADB debugging, and Debug over Bluetooth options on your *Android Wear* device, the next logical step is to get your development machine, which has the Android SDK environment installed, connected with your *Android Wear* device for debugging.

6.8.1 Wear Debugging via USB

Most *Android Wear* devices have a micro-USB port, and if your Wear device has one, you can easily connect your development machine to your *Android Wear* device using a micro USB cable—just like how we typically connect our Android phone or tablet devices to our development machines.

Once your Wear device is connected via USB, issuing the command `adb devices` on your development machine's command line will help verify that your Wear device is detected by adb.

Figure 6-21A shows the output of the command `adb devices`, which shows the serial number of my Gear Live *Wear* device. The serial number should match the serial number of your *Wear* device listed under your device's *Settings*.

```
$ adb devices
List of devices attached
R3AF700RH4E     device

$ █
```

Figure 6-21A adb devices—lists attached Android devices.

In case you do not see your *Wear* device listed, issuing the command `adb kill-server` will cause the adb server running on your development machine to stop. After that, the command `adb devices` will cause adb to start, with improved chances that you will see your *Wear* device listed. Figure 6-21B shows this sequence of commands.

```
$ adb kill-server
$ adb devices
* daemon not running. starting it now on port 5037 *
* daemon started successfully *
List of devices attached
R3AF700RH4E     device
```

Figure 6-21B adb devices—after adb "restart."

In case adb does not detect your *Wear* device, unplug the USB cable from the development machine side and run the command `tail -f /var/log/syslog /var/log/kern.log` and then plug the USB cable back into your development machine's USB port. Right after that, in a separate command window, run the `dmesg` command. You should see some output indicating the status of the USB connection.

The relevant output of the tail and dmesg commands are listed in the two snippets below:

```
==> /var/log/kern.log <==
Nov 29 13:28:44 acer-ubuntu13 kernel: [20896.774867] usb 3-1: Product: Gear Live
Nov 29 13:28:44 acer-ubuntu13 kernel: [20896.774872] usb 3-1: Manufacturer: Samsung
Nov 29 13:28:44 acer-ubuntu13 kernel: [20896.774876] usb 3-1: SerialNumber: R3AF700RH4E
==> /var/log/syslog <==
Nov 29 13:28:44 acer-ubuntu13 kernel: [20896.774867] usb 3-1: Product: Gear Live
Nov 29 13:28:44 acer-ubuntu13 kernel: [20896.774872] usb 3-1: Manufacturer: Samsung
Nov 29 13:28:44 acer-ubuntu13 kernel: [20896.774876] usb 3-1: SerialNumber: R3AF700RH4E
```

```
// Output of dmesg, to be run right after plugging in your Wear device
[21000.160893] usb 3-1: new high-speed USB device number 14 using xhci_hcd
[21000.177673] usb 3-1: New USB device found, idVendor=18d1, idProduct=d002
[21000.177679] usb 3-1: New USB device strings: Mfr=1, Product=2, SerialNumber=3
[21000.177682] usb 3-1: Product: Gear Live
[21000.177685] usb 3-1: Manufacturer: Samsung
[21000.177687] usb 3-1: SerialNumber: R3AF700RH4E
```

Another Linux command that can be useful for USB-related debugging is the `lsusb` command, which lists the USB devices connected to your development machine. If ADB is unable to see your *Wear* device, you will need to determine whether it is detected at the lower/operating system level. This will help in pinpointing and resolving the issue. The same concept applies on any operating system, which has their specific low-level tools that help investigate any issues, should they arise.

Figure 6-21C shows the output of the `lsusb` command, with the listing of my *Wear* device highlighted. It shows up with vendor ID 18d1 and vendor name Google, Inc. This will likely be true for all *Android Wear* devices because it's the adb driver that engages in order to manage the interface for the USB device. On Linux systems, the local

```
$ lsusb
Bus 002 Device 002: ID 8087:0024 Intel Corp. Integrated Rate Matching Hub
Bus 002 Device 001: ID 1d6b:0002 Linux Foundation 2.0 root hub
Bus 001 Device 002: ID 8087:0024 Intel Corp. Integrated Rate Matching Hub
Bus 001 Device 001: ID 1d6b:0002 Linux Foundation 2.0 root hub
Bus 004 Device 001: ID 1d6b:0003 Linux Foundation 3.0 root hub
Bus 003 Device 003: ID 064e:e330 Suyin Corp.
Bus 003 Device 002: ID 04f3:0023 Elan Microelectronics Corp.
Bus 003 Device 015: ID 18d1:d002 Google Inc.
Bus 003 Device 001: ID 1d6b:0002 Linux Foundation 2.0 root hub
$
```

Figure 6-21C `lsusb` command.

file `/usr/share/hwdata/usb.ids` lists the known USB vendor and device, IDs, and names included therein. You should see an entry for *18d1 Google, Inc.* in this file on your local Linux development machine. In case you are missing the `lsusb` command, the command `sudo apt-get install usbutils` will install USB utilities including the `lsusb` command.

```
shell@sprat:/ $ wm size
Physical size: 320x320
shell@sprat:/ $ wm density
Physical density: 240
shell@sprat:/ $ █
```

Figure 6-21D *wm* command.

Once you have `adb` connectivity established, you will be also able to shell into your *Wear* device by using the command `adb shell`. Once you are logged on to your *Wear* device, you can check out the kernel version info by using the command `cat/proc/version`. You can also query the device's size and density using the *wm* command (Figure 6-21D). The *wm* command can also be used to emulate a density and size that is different from its density and size. In this example, we are merely querying these properties.

```
$ adb shell
shell@sprat:/ $ cat /proc/version
Linux version 3.10.0-g20d4669 (android-build@vpbs1.mtv.corp.goo
gle.com) (gcc version 4.7 (GCC) ) #1 SMP PREEMPT Fri Oct 3 17:1
3:48 UTC 2014
shell@sprat:/ $ wm size
Physical size: 320x320
shell@sprat:/ $ wm density
Physical density: 240
shell@sprat:/ $ exit
$ █
```

Figure 6-21E Adb shell session.

Figure 6-21E shows the commands executed (and their outputs) on the *Wear* device.

While you can *adb shell* into the device and execute commands after that, you might find it convenient at times to execute commands on the *Wear* device from your development machine using the syntax `adb shell <command>`. The output of the command will become available on your development machine for storage or processing in one shot.

Figure 6-21F shows a command executed using the adb shell <command> approach. `uptime` is a common Linux command that tells you how long a host/device has been up since booting up.

You can also access the Android logging system using the adb logcat command.

```
$ adb shell uptime
up time: 08:27:29, idle time: 01:14:38, sleep time: 07:10:13
$ █
```

Figure 6-21F adb shell-based command executed directly from development machine.

```
$ adb  logcat | head -8
--------- beginning of system
I/Vold    ( 159): Vold 2.1 (the revenge) firing up
I/SystemServer( 456): Entered the Android system server!
I/SystemServiceManager( 456): Starting com.android.server.pm.Installer
I/Installer( 456): Waiting for installd to be ready.
I/Installer( 456): connecting...
I/SystemServiceManager( 456): Starting com.android.server.power.PowerMana
gerService
I/SystemServiceManager( 456): Starting com.android.server.am.ActivityMana
gerService$Lifecycle
$ █
```

Figure 6-21G adb logcat.

Figure 6-21G shows the output of the adb logcat command.

Thus, connecting your *Wear* device directly to your development machine via a USB cable is the simpler and easier way to get debugging setup on your device. The Bluetooth-based debugging option, which we will cover in the next section, happens to entail a USB cable-based connection with your handheld device and many more steps and effort to set up.

6.8.2 Wear Debugging via Bluetooth

Some Android Wear devices may come with wireless charging, and in any case, it is possible that some Android Wear device models lack a USB port. In case your *Android Wear* device lacks a USB port, getting debugging over Bluetooth will become essential. Because your *Wear* device is paired and tethered to your handheld Android device, you can debug your *Wear* device by connecting your development machine via USB to your handheld device and using your handheld device as an intermediary between your development machine and your *Wear* device.

First of all, if you had been successful or even attempted to get ADB working over USB directly between your development machine and your *Wear* device—covered in the earlier section—it would be best to reboot your *Wear* device and issue the command `adb kill-server` on your development machine. Because the following steps are quite elaborate and sensitive, it's ideal to start form a clean slate of a freshly started *adb* instance.

You will need to ensure that your *Wear* device has ADB Debug and Debugging Bluetooth enabled within *Settings → Developer options*. (This should already be the case if you kept up with the steps as described in Section 6.7.3.3 "Enabling Developer Mode and Debugging Settings on your *Wear* device.")

You will also need to ensure that your handheld device has Developer options and USB debugging enabled—this will likely already be the case, if you have been actively using your handheld Android device for development (Figure 6-22A).

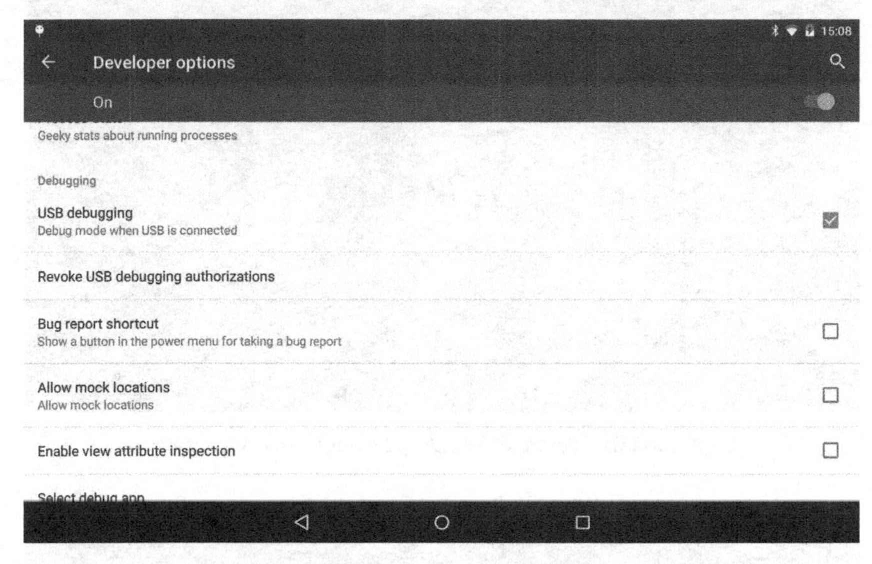

Figure 6-22A USB Debugging enabled on handheld device.

Next, you will need to verify that your handheld device is paired and tethered to your *Wear* device. You can do this by starting the *Android Wear* App on your handheld device and verifying that the *Wear* device is in a connected state.

Figure 6-22B shows the *Android Wear* App's screen upon starting it. Firstly, you will need to confirm that your *Android Wear* device is in a connected state. The *Wear* device name and the state of the connection are displayed on the top left of the screen shown in the figure. Next, you will need to access this App's *Settings* via the gear icons on the top right of the action bar.

Figure 6-22C shows the subitems under the *Android Wear* App's *Settings*. You will find that the default state of *Debugging over Bluetooth* item at the bottom is in an *off* state. You will need to enable this option.

Figure 6-22D shows the *Debugging over Bluetooth* option—soon after turning this option on, you will find certain connectivity status about the *Host* and *Target* show up. The Target represents your *Wear* device, while the Host represents your development machine.

Finally, you will need to execute a few commands on the command-line shell on your development machine as shown below:

```
adb forward tcp:<port - any port number greater that 1000> localabstract:/adb-hub
adb connect localhost:<same port number as above>
adb devices

adb -s localhost:<same port number as above>
```

Figure 6-22B Android Wear App *Settings*, connected state.

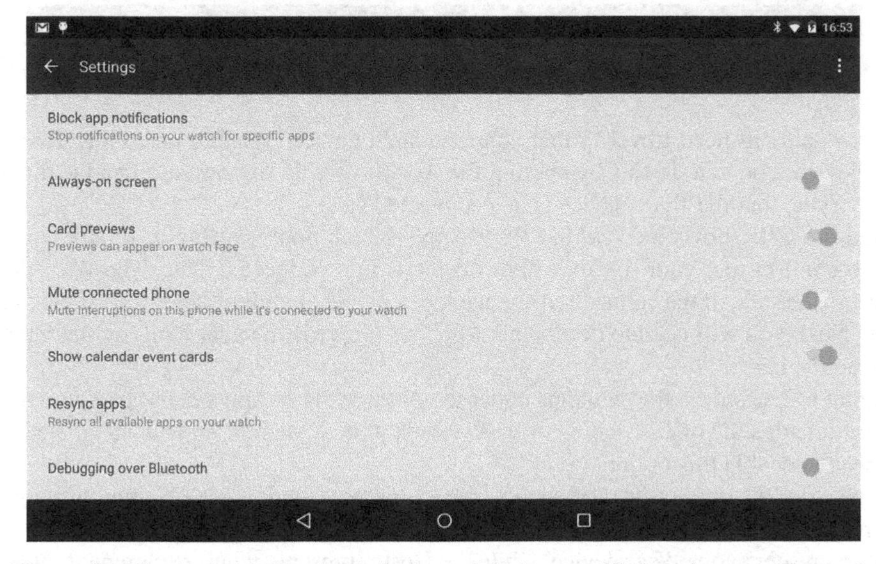

Figure 6-22C Android Wear App's Settings, default state of Debugging over Bluetooth.

adb port forwarding is used to set up an arbitrary port on localhost that forward requests to an abstract endpoint. This sets up a TCP/IP listener port on the local development machine, which forward requests via the intermediate handheld device to the *Wear* device.

Figure 6-22E shows the sequence of commands that I used in order to get Bluetooth-based debugging setup.

Once set up, you should be able to shell into your *Wear* device over adb.

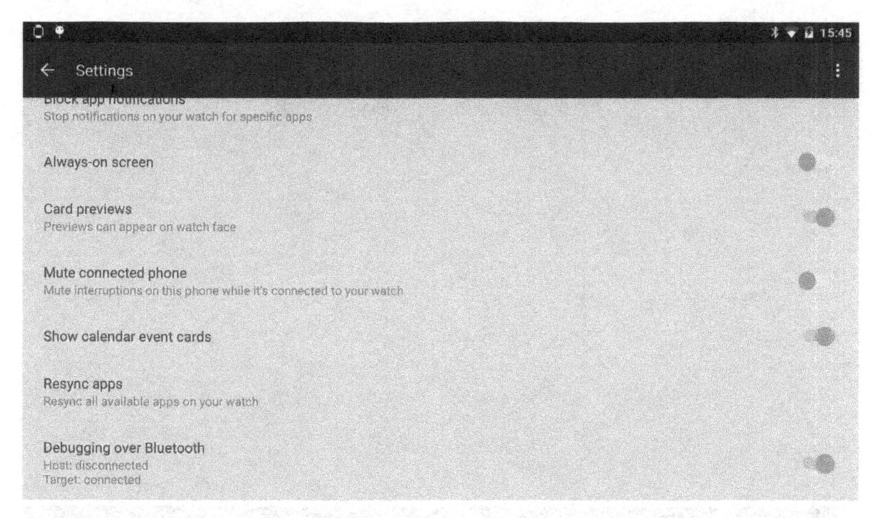

Figure 6-22D Android Wear App's Settings, enabling Debugging over Bluetooth.

```
$ adb kill-server
$ adb forward tcp:9729 localabstract:/adb-hub
* daemon not running. starting it now on port 5037 *
* daemon started successfully *
$ adb connect localhost:9729
connected to localhost:9729
$ adb devices
List of devices attached
00ea1b9d        device
localhost:9729  device

$ adb -s localhost:9729 shell
shell@sprat:/ $
```

Figure 6-22E adb commands to get Bluetooth debugging setup.

Also, once *Bluetooth forwarding* is set up, you will see that both the *Host* and *Target* show the connected status in the *Android Wear* App *Settings* under *Debugging over Bluetooth* (Figure 6-22F).

6.9 Peeking under the hood of your Wear Device

Now that we have been able to connect from the Development machine to the *Wear* device, let us take a quick peek under the covers.

Figure 6-23 shows the *zygote* and its child processes running on my *Wear* device. As we discussed in Chapter 3 on Android fundamentals, all Android Apps are run in virtual machine processes that are spawned by zygote. Therefore, taking a look at processes whose

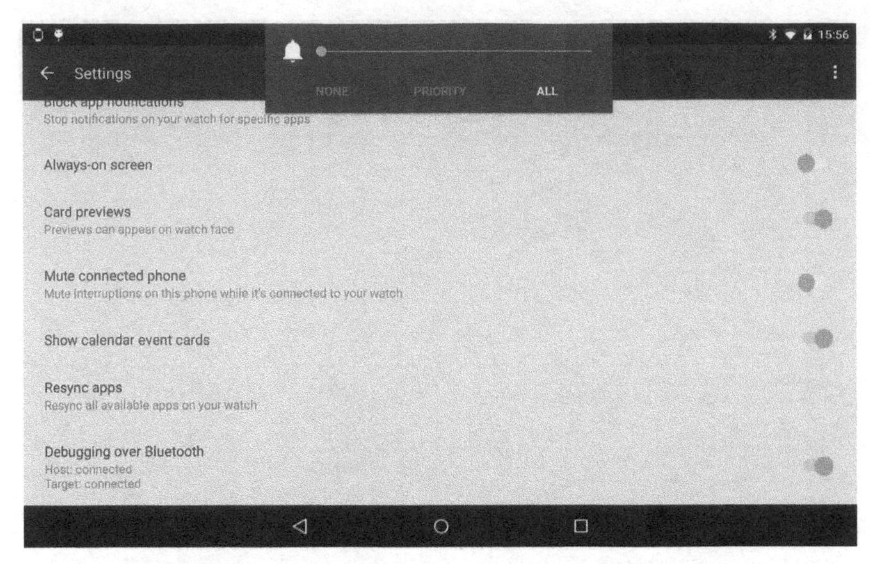

Figure 6-22F Host and Target both showing a connected status.

```
$ adb shell ps | grep zygote
root      163   1     297612 35552 ffffffff 00000000 S zygote
$ adb shell ps | grep 163
root      163   1     297612 35552 ffffffff 00000000 S zygote
system    456   163   366016 45160 ffffffff 00000000 S system_server
u0_a9     514   163   309280 22760 ffffffff 00000000 S android.process.media
u0_a3     574   163   360680 50256 ffffffff 00000000 S com.google.android.wearable.app
u0_a10    586   163   329492 29640 ffffffff 00000000 S com.google.android.gms
u0_a10    602   163   335068 31856 ffffffff 00000000 S com.google.process.gapps
u0_a10    614   163   336272 27128 ffffffff 00000000 S com.google.android.gms.wearable
u0_a10    671   163   322060 21460 ffffffff 00000000 S com.google.process.location
system    683   163   313200 19608 ffffffff 00000000 S com.google.android.apps.wearable.settings
u0_a12    733   163   326036 30568 ffffffff 00000000 S com.sec.android.wearable.watchface
bluetooth 794   163   328380 21228 ffffffff 00000000 S com.android.bluetooth
u0_a1     840   163   306460 20552 ffffffff 00000000 S com.android.providers.calendar
u0_a14    893   163   326452 29088 ffffffff 00000000 S com.google.android.apps.fitness
u0_a20    964   163   310432 19412 ffffffff 00000000 S com.texasgamer.tockle
u0_a7     1183  163   308456 22796 ffffffff 00000000 S android.process.acore
u0_a8     1214  163   306592 17480 ffffffff 00000000 S com.android.defcontainer
u0_a22    1600  163   311932 21508 ffffffff 00000000 S com.google.android.music
$
```

Figure 6-23 Wear device zygote child processes.

parent process ID is the zygote process' process ID (PID 0) tells us what Apps are running on the device. I notice that there is an App running whose package name is *com.google.android.wearable.app*. This is the package identical to the *Android Wear* app that we installed from the Play Store earlier:

https://play.google.com/store/apps/details?id=com.google.android.wearable.app

This is consistent with the expected behavior wherein the handheld "companion" App pushes and installs the *Wear* version of the App onto the *Wear* device after pairing and tethering. I also notice that the Tockle App (*com.texasgamer.tockle*)—which I happened to have installed from the Play Store on my handheld device—shows up as a running App on my *Wear* device. The Tockle App enables control of your phone from your *Wear* device including toggling your Wi-Fi connectivity and other system settings.

6.10 Engaging your Android Wear device via Notifications

In the simplest scenario, the user can choose to sync Notifications to the *Android Wear* device so that Notifications that are displayed on the handheld device are also displayed on the *Wear* device over the tethered connection. Android version Lollipop (version 5) clears Notifications on all devices when the user clears it on one device. Thus, Notifications that are synched to *Wear* devices help engage users with their *Wear* device, and this represents the simplest entry-level engagement of users with their *Wear* devices. Depending on the needs of your App, you can engage your App with *Wear* devices in various ways, at different levels of engagement, and these have been covered in this section.

6.10.1 Engaging Android Wear via Notification Sync

Because Notifications can be synced between a user's handheld device and *Wear* device, at the simplest level of engagement, Notifications are presented on the *Wear* device as well. This is the simplest form of engagement with *Wear* devices.

6.10.2 Wear Extended Notifications

The next progressive degree of engagement with *Wear* entails designing Apps that extend Notifications to exhibit *Wear*-specific user interfaces and behavior. The Android v4 support library provides the *NotificationCompat.WearableExtender* class, which supports adding *Wear* extensions to Notifications. Using the *WearableExtender*, you can provide actions that are specific to *Wear*. Thus, Notifications can have a different actions and user experience across the handheld and *Wear* platforms. On the *Wear* devices, we need to emphasize simplicity of interaction.

A quick review of the interaction flow, when receiving a Notification from the Gmail App, provides a good example of how you might design your App's extended Notification.

Figure 6-24A Gmail arrival Notification.

Figure 6-24A shows the receipt of an email. Tapping on the Notification will display the content of an email (Figure 6-24B).

Figure 6-24B shows the display the content of an email. Swiping right will display the Archive action on the email item (Figure 6-24C).

Figure 6-24B Gmail reading an email.

Figure 6-24C shows the Archive action, and swiping right offers other actions—Reply (Figure 6-24D).

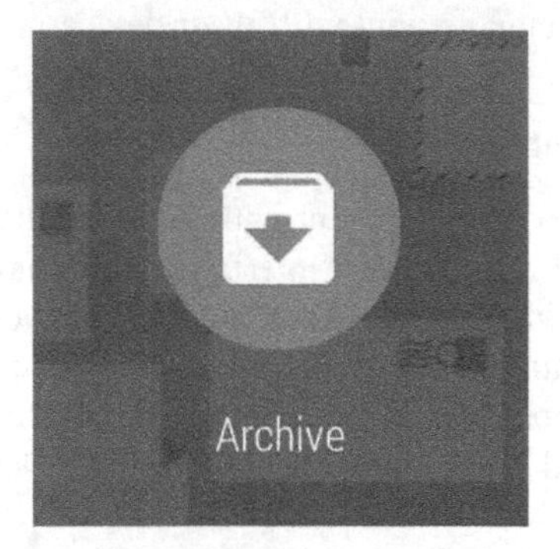

Figure 6-24C Archive.

Figure 6-24D shows the Reply action, and swiping right offers the last action—Open on handheld device (Figure 6-24E).

Figure 6-24D Reply.

Figure 6-24E shows you the option to open the email item on your phone or handheld device.

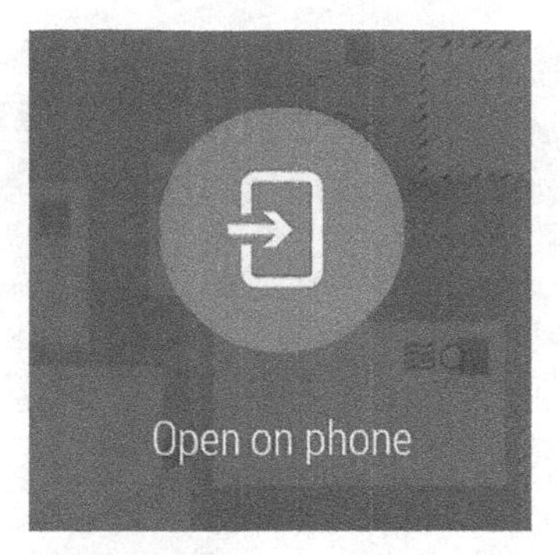

Figure 6-24E Open on phone.

Studying these above *Wear*-based user interfaces and interactions can provide you with ideas for designing your own *Wear* App.

6.11 Android Wear Targeted Apps

During initial development, *Android Wear* apps can be directly installed and executed on your development *Android Wear* device—your *Wear* App will need to be targeted specifically for the *Android Wear* platform whose API level is 20. An App that has been built for any of the handheld Android OS versions will but, naturally, fail to install onto an *Android Wear* device. In Development/Debug mode, you can directly build and install your *Wear* App's apk onto your *Wear* device. However, *Android Wear* devices that do not have the ability to access the Play Store, particularly the Play Store App, has been excluded from the *Android Wear* software stack. Therefore, for purposes of distribution—in order to release a production Wear App to consumers—you will need to build an Android handheld companion App and embed your *Wear* App within its *res/raw* directory. This is an intricacy that *Android Studio* handles for you when you select the target form factor to handheld devices as well as *Android Wear* (API level 20).

6.12 Hello Wear World: Writing our first Wear App

We will now cover the subject of writing, installing, and executing a simple *Wear* App on a *Wear* device or suitable AVD. Writing your first and simple hello *Wear* App is fairly easy once you have your environment set up and working properly. We will be using *Android Studio* to do so in this section. Once you open up *Android Studio*, you will need to create or *Start a new Android Studio project.*

Figure 6-25A shows the new project screen, which requires you to enter the name of your application, the company domain name (or package name in reverse order), and the project location in your local file system. The next screen helps you select and target the device form factor.

Figure 6-25B shows the new project's Form Factor selection screen, which helps you choose the form factors that the App is intended to run on. In this case, I selected solely *Android Wear* platform, because I intend this to be a *Wear* App, in development mode.

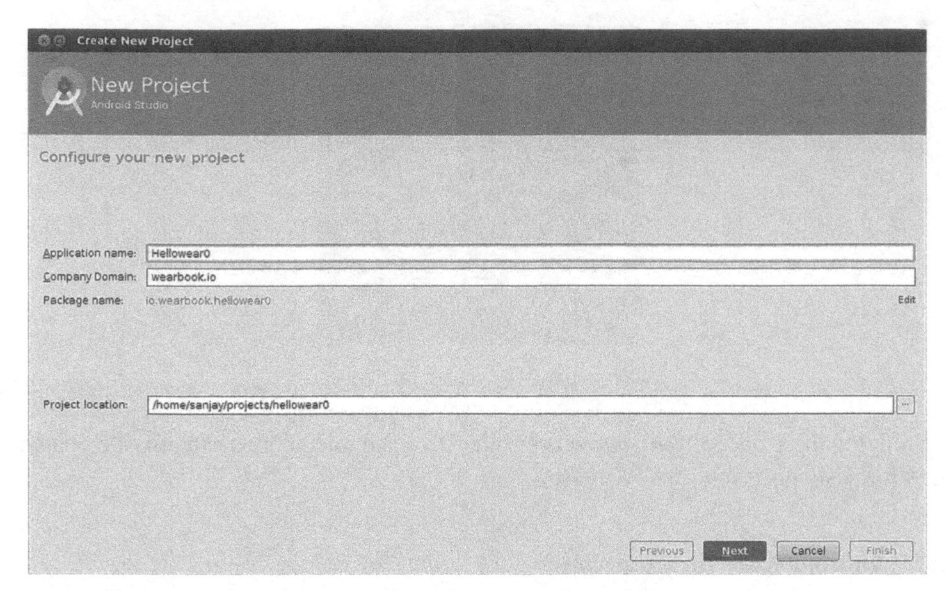

Figure 6-25A Android Wear new project.

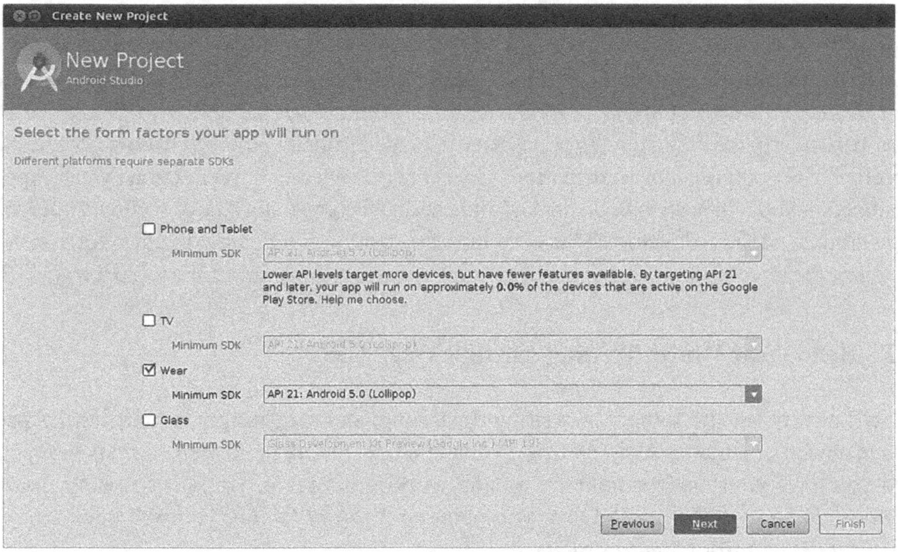

Figure 6-25B Android App, selecting form factor.

Figure 6-25C shows the new *Wear* project's code artifacts, which you can browse and edit. Figure 6-25D shows some trivial editing of the *Wear* project's code artifacts. Figure 6-25E the step of building your first *Wear* App by clicking on the green "play" button.

If you have your *Android Wear* device connected, you will be prompted to choose the device that your App will be installed on—as shown in Figure 6-25F.

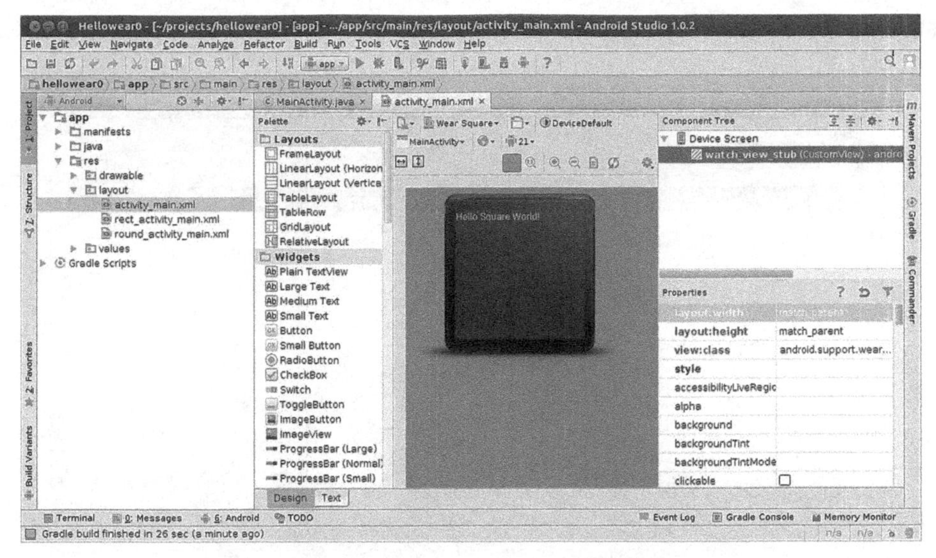

Figure 6-25C Android Wear project code artifacts.

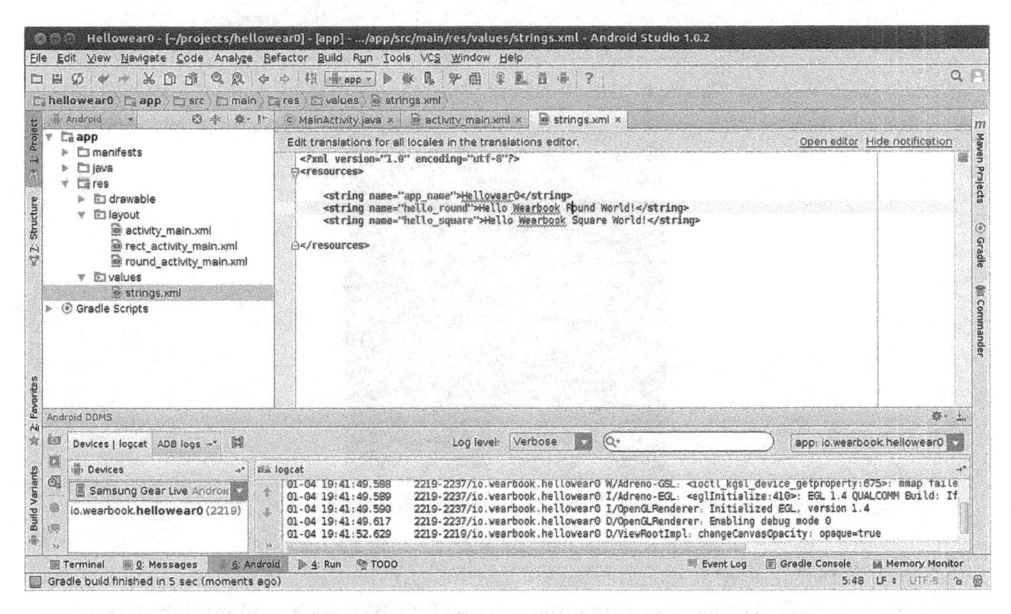

Figure 6-25D Android Wear, editing project code artifacts.

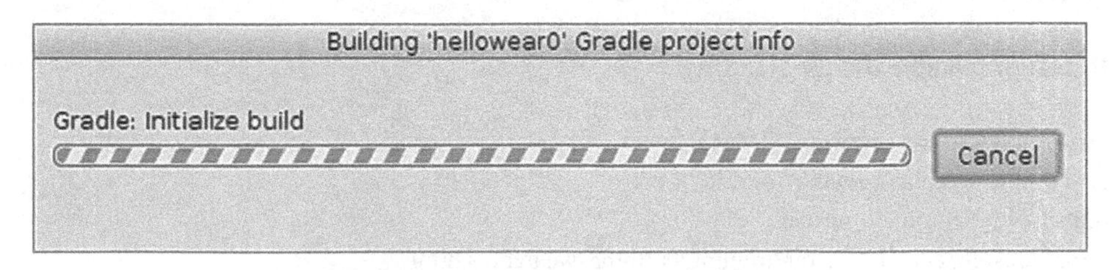

Figure 6-25E Android Wear, building App.

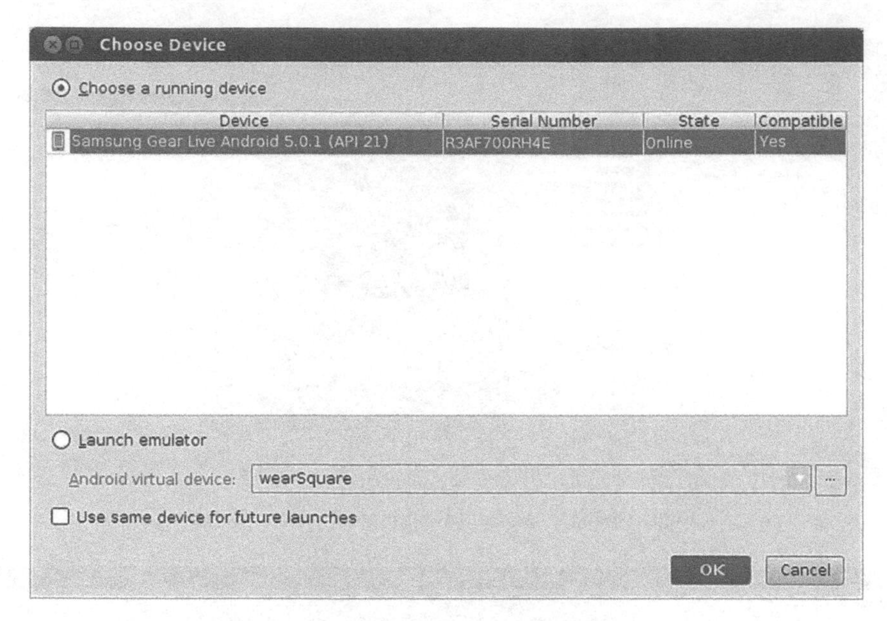

Figure 6-25F Android Wear, choose device.

Figure 6-25G Hello Wear App running on Android Wear device.

Figure 6-25G shows the Hello App running on the Android Wear device.

You may browse though the source tree of your first *Wear* App and get familiar with its project's artifacts.

References and Further Reading

http://en.wikipedia.org/wiki/WIMM_One
http://en.wikipedia.org/wiki/Android_Wear
http://www.android.com/wear
https://developer.android.com/training/building-wearables.html
https://developer.android.com/reference/packages-wearable-support.html
https://support.google.com/androidwear/answer/6056843?hl=en

Chapter 7 Android Wear API

7.1 Google Services and Google Play Services

Google Services include a wide range of Google-powered features and functionalities such as Maps, Location, Drive, Wallet, Games, Google+, Google Cloud Messaging, *Android Wear*, *Google Fit*, authentication, and more. Information on Google Services is available at http://developer.android.com/google/index.html.

Both the *Android Wear API* and the *Google Fit API* happen to be a part of the *Google Play Services* and even though *Google Play Services* provides a wide range of APIs not limited to *Android Wear* or *Google Fit*, we will begin this chapter by covering *Google Play Services*.

Google Play Services helps Apps avail of the latest Google-powered features. *Google Play Services* also provides system level features such as a dynamic security provider, which provides an OpenSSL implementation that can be updated dynamically, thus helping Apps benefit from timely security patches independently from the OS updates. Google Play Services are available on most Android devices, but is not part of the Android OS.

There is a Google Play Services App available on the Google Play Store at:

https://play.google.com/store/app/details?id=com.google.android.gsm

Once installed on the user's device, the *Google Play Services* App runs various services on the Android device some of which work closely with corresponding Google services on the cloud. You will notice that the Google Play Services App has a package name of *com.google.android.gsm*. "GSM" is generally considered to be an acronym for Google Services for Mobile.

Apps running on a device can connect as clients to these services running on the device and communicate via Android's IPC mechanisms. You may have noticed the Google Play

Wearable Android™: Android Wear & Google Fit App Development, First Edition. Sanjay M. Mishra.
© 2015 John Wiley & Sons, Inc. Published 2015 by John Wiley & Sons, Inc.

Services App's services running on your device when accessing *Running* Apps under your device's *Settings*. The Google Play Services App has an update cycle and versioning that is mostly independent of the Android OS update cycle and versioning. At the time of writing, the current version of Google Play Services is at *6.5* per the versioning scheme for the App that is visible to the end users (Figure 7-1A).

Figure 7-1A Google Play Services running on a handheld device.

Your App will need to depend on the *Google Play Services* library project in order to connect to the Google Play Services App that is running on the device in order to avail of particular features/APIs.

The *Google Play Services library project*, which is distinct from, and represents the client side of, the Google Play Services App, can be found under your Android SDK installation home at

<ANDROID_HOME>/extras/google/google_play_services/libproject.

This library uses an integer-based version numbering scheme, which at the time of writing is at *6587000*. This differs from but is related to the *6.5* version scheme that users can see.

```
$ cat /opt/androidsdk/extras/google/google_play_services/libproject/google-play-serv
ices_lib/res/values/version.xml
<?xml version="1.0" encoding="utf-8"?>
<resources>
    <integer name="google_play_services_version">6587000</integer>
</resources>$
$ 
```

Figure 7-1B Google Play Services library project version.

Figure 7-1B shows the library version of Google Play Services library project (*6587000*), which is the current version at the time of writing.

The Google Play Services App includes Android Services running on your device. The Google Play Services library helps your App connect to the Google Play Services' Android

Services running on your device. Your App connects to the Google Play Services via the library over AIDL/IPC. The Google Play Services App that is running on your device can act as a proxy for the Google functionality on the cloud.

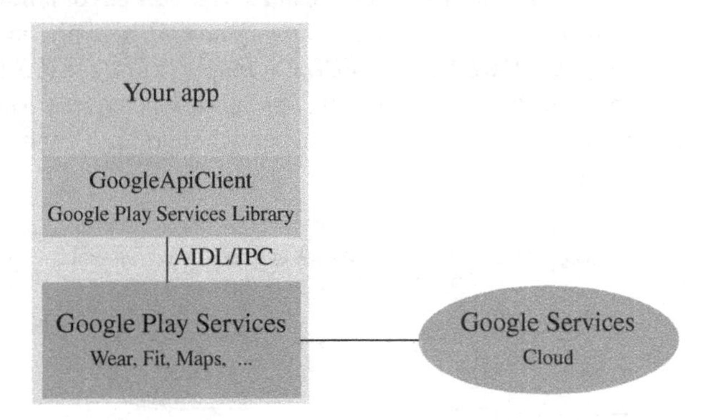

Figure 7-2 Google Play Services and its relationships with Google Play Services library.

Figure 7-2 shows the relationship and interaction between your App, the Google Play Service client library, the Google Play Services' running Android Service components, and the Google Services on the Cloud.

7.1.1 GoogleApiClient class

The *GoogleApiClient* is the entry point that enables your Apps to make a connection to the Google Play Services and thereafter avail of particular features/APIs (such as Location, Maps, Drive, Wallet, Games, *Android Wear*, Google Fit, and so on). The *GoogleApiClient* class provides you with the ability to connect to Google Play Services and perform synchronous and asynchronous calls to any of the various particular services.

Figure 7-3A shows an overview and partial listing of the *GoogleApiClient*, which resides in the package *com.google.android.gms.common.* The nested *GoogleApiClient.Builder*

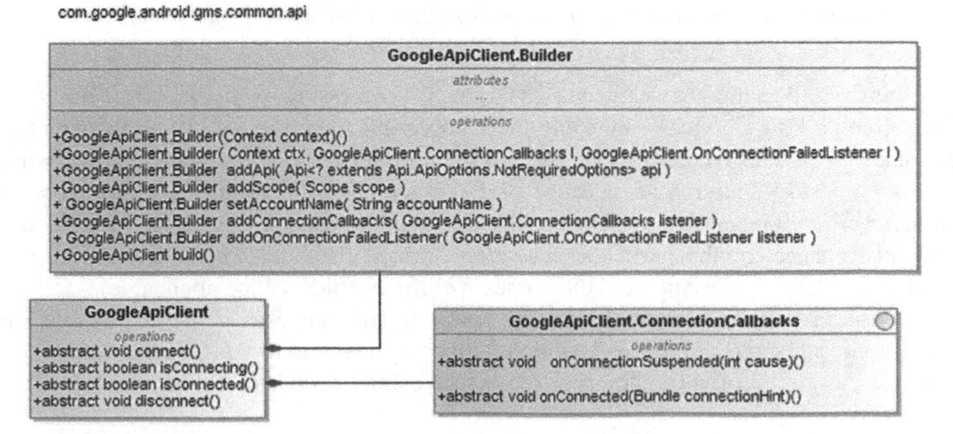

Figure 7-3A GoogleApiClient class diagram, partial listing.

class is used to set up the needed API, specify the scope, and register the various connection callback listeners before *build()*-ing the *GoogleApiClient* instance.

Other related classes in the same package include the *ConnectionResult* class, which is not shown in the diagram. The *ConnectionResult* encapsulates the success or failure of the attempt at connecting to the Google Play Services. The *ConnectionResult* also provides useful statues and constants such as *API_UNAVAILABLE, SIGN_IN_REQUIRED, DEVELOPER_ERROR*, and so on, which can be used in the *GoogleApiClient.OnConnectionFailedListener* implementation, to determine the reason for the failure when attempting to connect to the Google Play Services from your App.

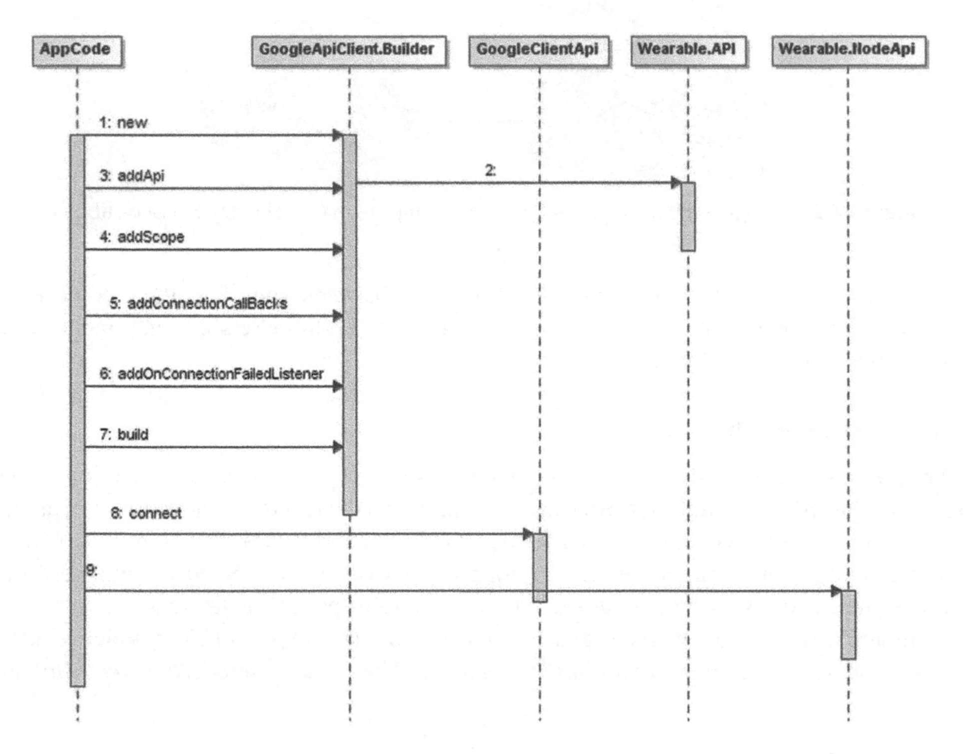

Figure 7-3B Sequence diagram GoogleApiClient, accessing particular feature/API.

Figure 7-3B shows a sequence diagram for accessing a particular feature/API from Google Play Services—by using the *GoogleApiClient.Builder*; the Builder pattern pervades the various Android APIs. The call to *addApi()* adds the particular feature, while the call to *build()* returns an instance of *GoogleApiClient.* Invoking *connect()* on the *GoogleApiClient* instance connects it to the Google Play Services. After the client is successfully connected, the particular feature-specific APIs that were added earlier will become available to the App (and the scopes specified earlier will be applicable).

In case an App needs multiple features from Google Play Services, it is possible—in general—to add multiple APIs and multiple scopes to the same *GoogleApiClient* by making multiple calls to *addApi()* and *addScope()* before calling *connect()*. However, particularly for accessing the Wear API using the *GoogleApiClient*, it is recommended to use a separate instance of the *GoogleApiClient* exclusively for accessing the Wear API via the *Wearable* class (covered in the next section). The *GoogleApiClient*'s attempt to connect will fail if the *Android Wear* App is not installed on the device. Appending multiple APIs that include the Wear API and other APIs into one *GoogleApiClient* instance can render the other APIs to become inaccessible, although they are available.

Detailed documentation on Google Play Services is available at

https://developer.android.com/google/play-services/index.html and

http://developer.android.com/reference/com/google/android/gms/common/package-summary.html.

In the year 2014, five versions of Google Play Services (in the series 4.x, 5.x, and 6.x) were released, along with the introduction of new constructs and updated best practices as well as deprecation of some classes and interfaces. These changes are generally accompanied by useful documentation and sample code. If you are using Google Play Services in your App, it is important to keep your App abreast with the latest version of the library and the associated best practices.

7.2 Android Wear Network

At the center of the *Android Wear* platform and API lies the concept of the *Android Wear Network*, which is a network of devices or "nodes." The devices on the *Android Wear Network* include *Wear* devices and other devices (such as handheld devices) that the Wear devices connect to and interact with.

In a world of a multitude of devices, there is a dynamic landscape of devices on this *Wear Network* that appears and disappears over time, as devices join and drop off this network. While servers and computers on a stable wired network are set in a relatively static landscape, wearable devices (which categorize as a body area network) tend to join and drop off the wear network more dynamically due to factors such as their fluctuating proximity to other devices and networks, loss of battery power, and so on. If you happen to walk away from your desk on which your Android handheld device is placed, while wearing your smart watch tethered to it, the connectivity between your handheld device and your wear device is liable to undergo a disruption after some point. In such a backdrop, a given node may have the need to detect other nodes, send messages to them, or sync some data across them—in a secure manner that respects the boundaries and separateness of individual Apps and their data.

7.3 Android Wear API, in depth

The Wear API contains at a high level a Node API, Data API, and Message API. The package documentation is available at

https://developer.android.com/reference/com/google/android/gms/wearable/package-summary.html.

7.3.1 Wear API: wearable package

The *Android Wear* API's main package, *wearable* (*com.google.android.gms.wearable*), contains about seven interfaces and 10 classes.

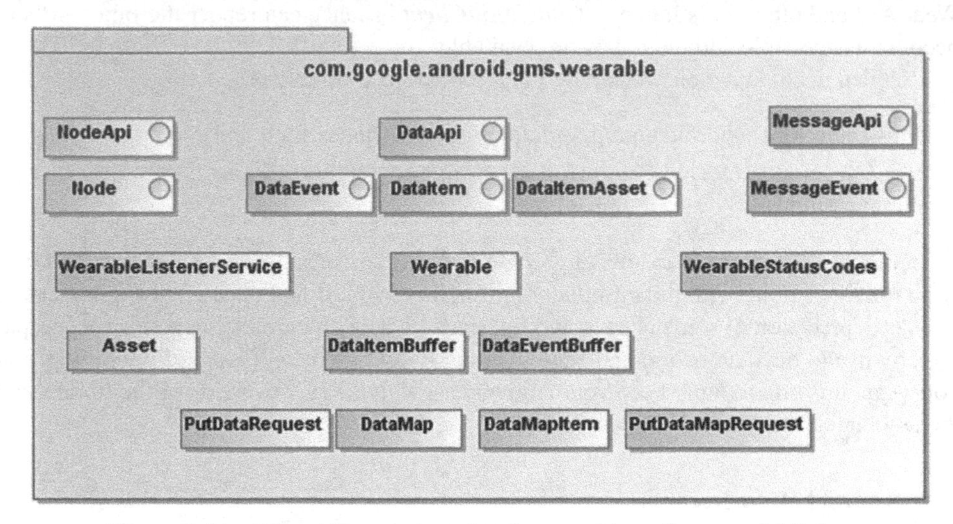

Figure 7-4A Wear API, overview of main *wearable* package, partial listing.

Figure 7-4A shows an overview of the interfaces and classes within the *wearable* package, which includes the *Wearable* class and key interfaces *NodeApi*, *DataApi*, and *MessageApi*. The *NodeApi* supports Apps in becoming aware when nodes join and leave the Wear Network. The *DataApi* helps read, write, and sync App data between nodes. The *MessageApi* helps Apps send transient messages between nodes.

7.3.1.1 Node interface The *Node* encapsulates basic information about a device or host on the *Android Wear Network.*

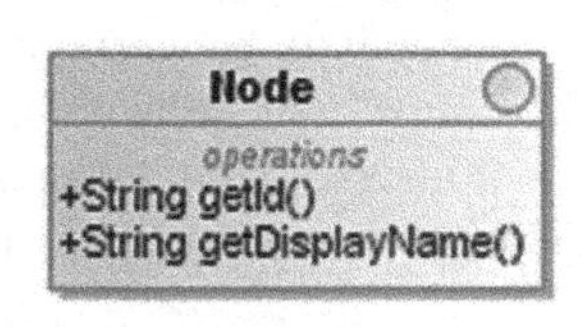

Figure 7-4B Node interface, complete listing.

Figure 7-4B shows the *Node* interface diagram. Conceptually, the *Node* has two attributes—the display or human readable name (often generated from the Bluetooth device name) and an identifier string.

7.3.1.2 WearableListenerService Another key class is the *WearableListenerService* that supports Apps in receiving node connectivity events and other events via the *DataApi* and *MessageApi*, which are covered in detail in the next sections.

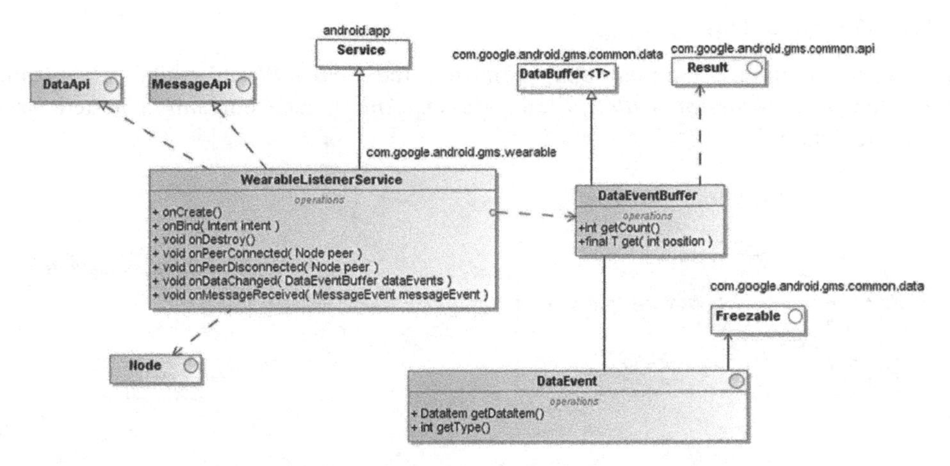

Figure 7-4C WearableListenerService, partial listing.

Figure 7-4C shows the class diagram for the *WearableListenerService*, which extends *android.app.Service.* Applications that are interested in receiving node events or data/message events via the *DataApi* and *MessageApi* while in the background will need to extend this class. There can be only one *WearableListenerService* class in your application. Both your Wear App and the companion handheld Apps will typically implement their own *WearableListenerService* if each is interested in keeping updated of changes in the *Android Wear* Network. As it turns out, the handheld device that's paired and tethered to the *Wear* device is also a node in the *Android Wear* Network.

7.3.1.3 DataEvent The *DataEvent* is an interface for receiving data changes via the data change listener in the *DataApi* or the *WearableListenerService*.

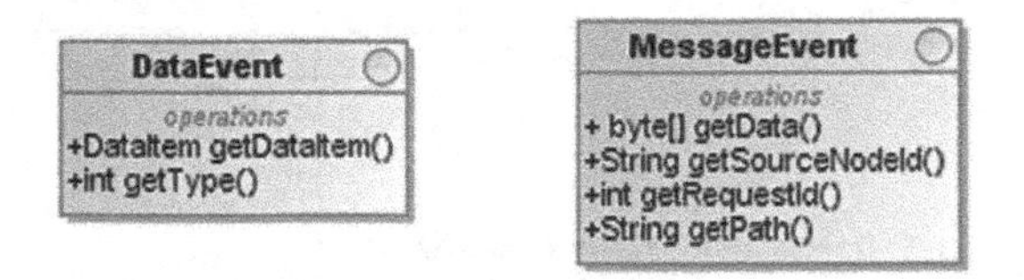

Figure 7-4D DataEvent and MessageEvent.

Figure 7-4D shows the *DataEvent* (as well as the *MessageEvent*, which is covered in the next section). The *getDataItem()* method provides access to the data item, while the *getType()* method specifies whether the data was changed(*TYPE_CHANGED*) or deleted(*TYPE_DELETED*).

7.3.1.4 MessageEvent The *MessageEvent* is an interface for receiving messages via *onMessageReceived()* in the *MessageApi* or the *WearableListenerService*. Figure 7-4D shows the MessageEvent (along with the *DataEvent* covered in the previous section). The *MessageEvent*'s *getData()* method returns the payload of the message.

7.3.2 Wearable class

The *Wearable* class is the main entry point into the *Wear API* and contains key static attributes: *API*, *NodeApi*, *DataApi*, and *MessageApi*. It also contains a nested class *WearableOptions*.

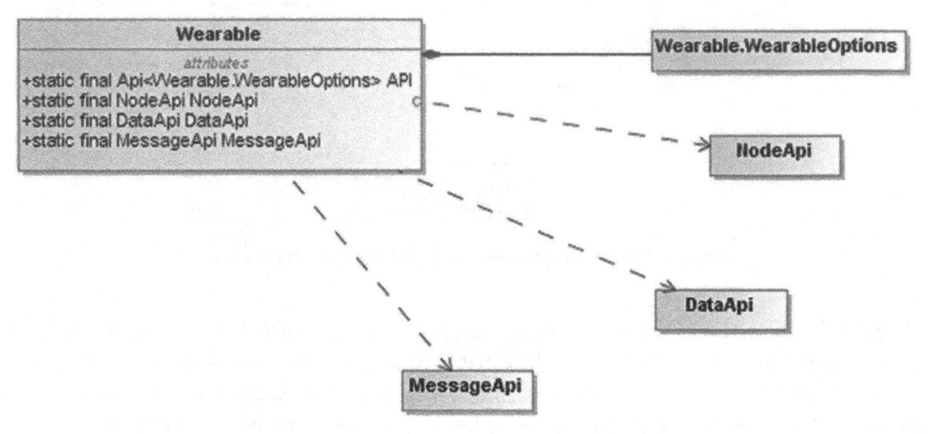

Figure 7-5A Wearable class diagram.

Figure 7-5A shows the Wearable class and its attributes and nested class *WearableOptions*. The static attribute *API* is used in invoking the *addApi()* method while building the *GoogleApiClient* instance.

The inner class *Wearable.WearableOptions* represents the API configuration parameters for the Wear API.

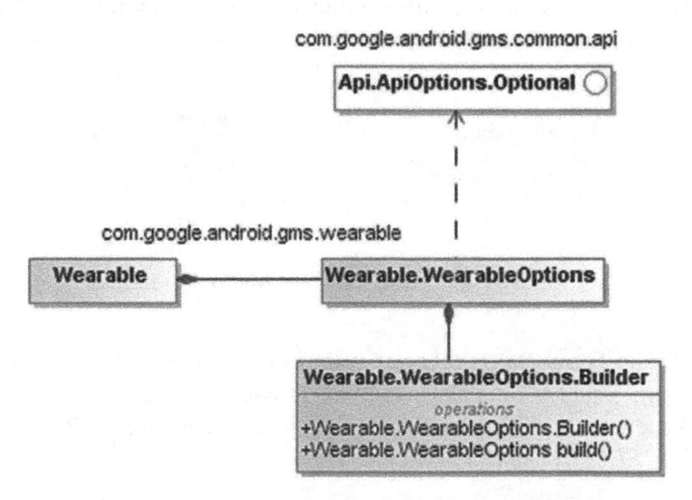

Figure 7-5B Wearable.WearableOptions inner class.

Figure 7-5B shows the *Wearable.WearableOptions* inner class, which implements the *Api. piOptions.ApiOptions* interface from the *com.google.android.gms.common.api* package.

7.3.3 NodeApi

The *NodeApi* interface supports detecting and learning about nodes as they connect and join the Wear network or disconnect and drop off. The *NodeApi* can be said to be tuned into the Node connection and disconnection events on the Wear network. When the *NodeApi* detects these Node connection and disconnection events, it delivers them to the Wear Apps, which have registered interest in these events.

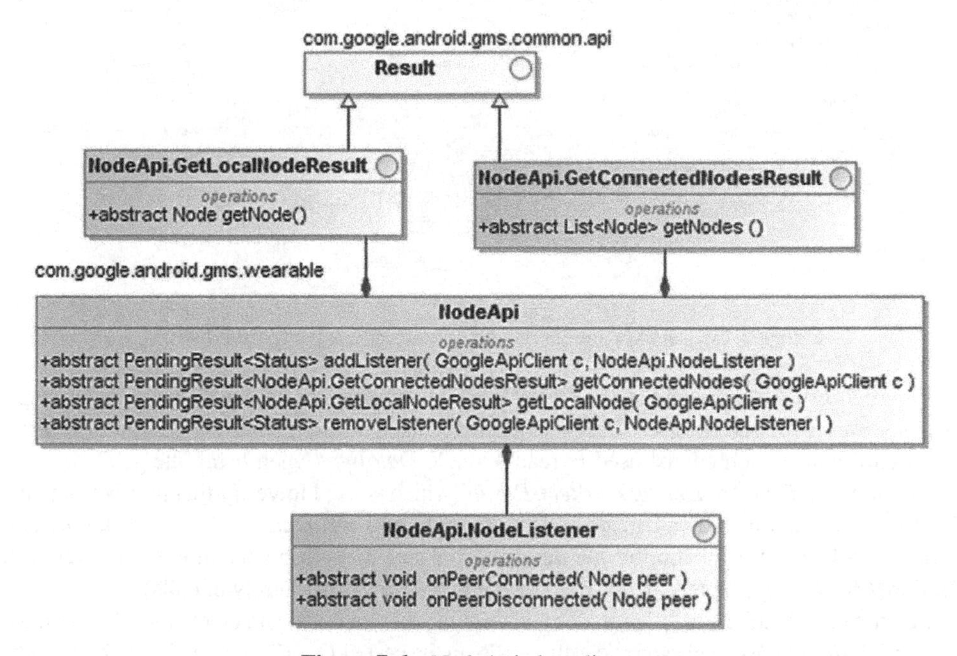

Figure 7-6 NodeApi class diagram.

Figure 7-6 shows the class diagram of the *NodeApi* class and it has the nested interface *NodeApi.NodeListener* (which helps your App detect that a peer on the *Wear Network* has connected or disconnected), *NodeApi.GetConnectedNodesResult* (which helps your App obtain a list of connected nodes on the *Wear Network*), and *NodeApi.GetLocalNodeResult* (which helps your App in accessing the node object which represents "this" device, the local device that your App is running on).

7.3.4 DataApi

The *DataApi* interface and its associated family of interfaces and classes provide Apps with the ability to read, write, and synchronize data across the devices on an *Android Wear* network while maintaining the privacy of the data within the App. Fundamentally, the *DataApi* provides the put and get functionality in order to read and write data to the Wear Network. The *DataApi* works closely with the *DataItem*, which encapsulates the data.

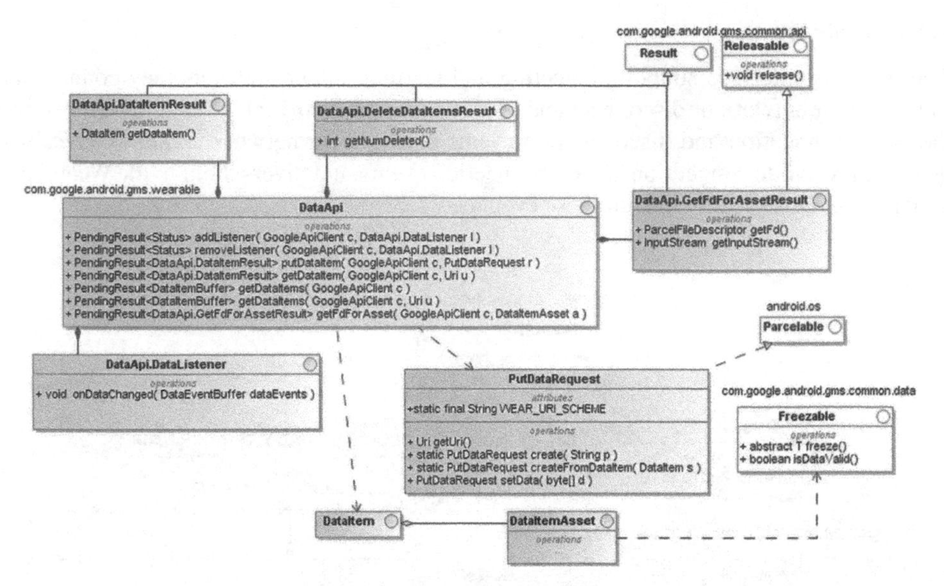

Figure 7-7 DataApi and its related interfaces, classes (partial listing).

Figure 7-7 shows a partial listing of the *DataApi* interface and its inner interfaces *DataApi.DataListener* (which can be used to register a listener using the *addListener()* method), *DataApi.DataItemResult* (which can be used to read a single *DataItem* when using the *getDataItem()* family of calls), *DataApi.DeleteDataItemsResult* (which is used to verify that number of items deleted upon making the call to *deleteDataItems()*), and the *DataApi.GetFdForAssetResult* (which can be used to obtain the file descriptor of an *Asset*—a binary blob data covered in Section 7.6.1 of this chapter—when making the *getFdForAsset()* family of calls).

Fundamentally, the *DataApi* provides the ability to add, retrieve, and delete *DataItems* on the *Android Wear Network*. It also provides the ability in the nested *DataListener* to listen for changes in the data. The *WearableListenerService*, which implements the *DataApi* and was covered earlier in this chapter, is useful for listening for data changes in the background. Calling the *getDataItems()* on booting can help keep up to date with changes while the device was offline.

7.4 DataItem, DataMapItem, and DataMap

DataItem, *DataItemMap*, and *DataMap* are closely related, and furthermore, they are all related to the *PutDataRequest* and *PutDataMapRequest*, which are covered in the very next section.

7.4.1 DataItem

A *DataItem* is an interface that represents the data that needs to be read and written via the *DataApi* to the *Android Wear* network. *DataItems* are inherently synchronized across all the nodes in the user's *Android Wear Network*. *DataItems* can be set on a local node even while that node is not connected to the Wear network. Whenever such a local node gets connected to the Wear network, the local *DataItems* that are pending synchronization will get synchronized.

Each *DataItem* contains a payload, which is a *byte* array. The payload is intended to be small in size—up to about 100 KB—but in practice you must strive to keep your *DataItem*'s payload much smaller than this limit. As with all remote procedure calls and interprocess communication, it is best to lean toward the side of being more conservative. Chapter 3 on Android IPC and AIDL has covered some of these constraints on the size of the payloads when performing IPC in general. *Assets*—covered in a subsequent section—are appropriate for encapsulating larger binary, blob data such as images. Each *DataItem* is identified by a path or URI on the Wear network in the format *wear://<node_id>/<path>*, where *node_id* represents the node that created the data, while the *path* portion of the URI is defined by the application.

A given *DataItem* remains private to the App that created it. Although the *DataItem* is a fundamental block for the persistence and synchronization of data on the Wear network, it does not have a constructor or Builder—the classes that are used when creating and using *DataItems* in the *DataApi* operations include the *PutDataRequest*, *DataMapItem*, and *DataMap*; these have been covered in the following sections.

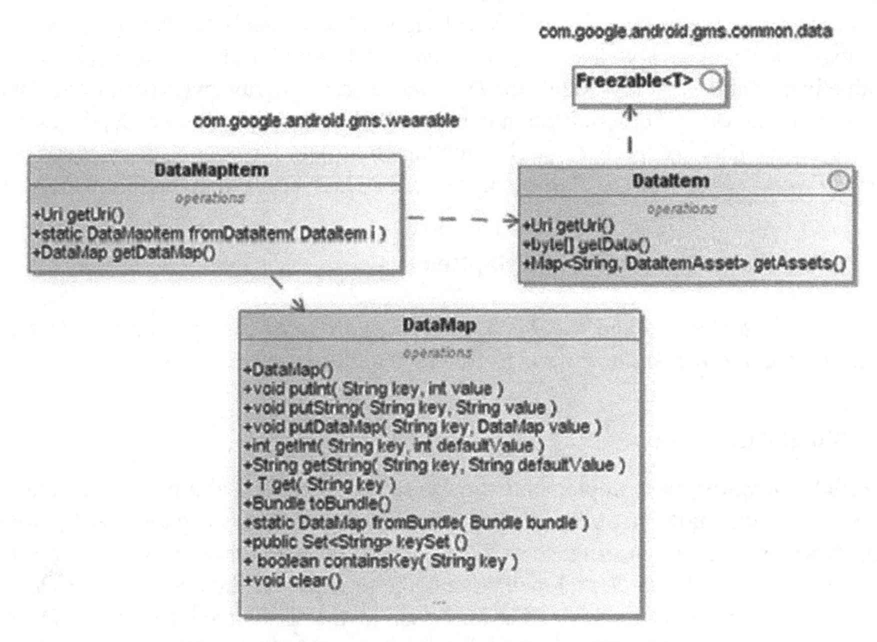

Figure 7-8 DataItem, DataItemMap, and DataMap.

Figure 7-8 shows the *DataItem* interfaces along with the *DataItemMap* and *DataMap* classes.

The *DataItem* interface extends the *Freezable* interface from the *com.google.android.gms.common.data* package. The *Freezable* interface pertains to transforming volatile data into an immutable representation via its *freeze()* method. When the application code receives data change events in the listener such as *onDataChanged()*, it receives a *DataEventBuffer* of data events. When items in this buffer are used outside of the call scope, *freeze()* should be invoked on such items.

DataItems are not intended for concurrent modifications and attempting so can result in inconsistent results. Therefore, it becomes the responsibility of the App to implement its concurrent modification strategy. In one approach, only the creator node that originally created the *DataItem* may modify the data. There are several other approaches, and it is left to the App to use a concurrent update strategy for its *DataItems*.

7.4.2 DataMapItem

The *DataMapItem* wraps a *DataItem* like object and is a structured and serializable representation of a *DataItem*. The original *DataItem* is not expected to be modified after the *DataMapItem* has been created based on it. Figure 7-8 in the previous section includes the class diagram of the *DataMapItem*.

The *DataMapItem* is closely associated with the *PutDataMapRequest* and the DataMap classes, both of which are covered in the following sections.

7.4.3 DataMap

The *DataMap* provides a key value pair-based support for storing basic data types such as int, long, float, double, byte, as well as arrays and *Bundles*. Figure 7-8 shows some of the methods available in the *DataMap* class. While the *DataItem* works with raw bytes of payload data and requires serialization and deserialization to be handled in the application code, the *DataMap* provides off-the-shelf support for storing fundamental data types as well as *Bundles*. The *DataMap* works along with the *PutDataMapRequest* class, covered in the following section.

7.5 PutDataRequest and PutDataMapRequest

Both the *PutDataRequest* and *PutDataMapRequest* are used for creating new *DataItems* and adding them to the *Android Wear* Network by calling the Data API.

7.5.1 PutDataRequest

The *PutDataRequest* class implements the *Parcelable* interface and is used to create and encapsulate new DataItems on the *Android Wear* Network. It has a static *create()* method which supports creating an instance of *PutDataRequest* with an associated, encapsulated *DataItem* instance. The *setData()* method sets the payload on the encapsulated *DataItem* instance. The *putAsset()* method adds an Asset to the associated *DataItem*. Once the *PutDataRequest* instance has been populated, it can be added to the *Android Wear Network* by calling the *DataApi*'s *putDataItem()* method.

Figure 7-9 shows class diagram of the *PutDataRequest* class that shows several of its available methods. The *PutDataRequest* is the key class that works with the *DataApi* in order to make the request to put/add *DataItems* to the *Android Wear Network*.

7.5.2 PutDataMapRequest

The *PutDataMapRequest* is a useful utility closely associated with and secondary to the *PutDataRequest*. The *PutDataMapRequest* and *DataMap* classes make it convenient to sync data to the *Android Wear Network* by using the serialization and de-serialization that's available off the shelf. The *PutDataMapRequest* encapsulates a DataMap.

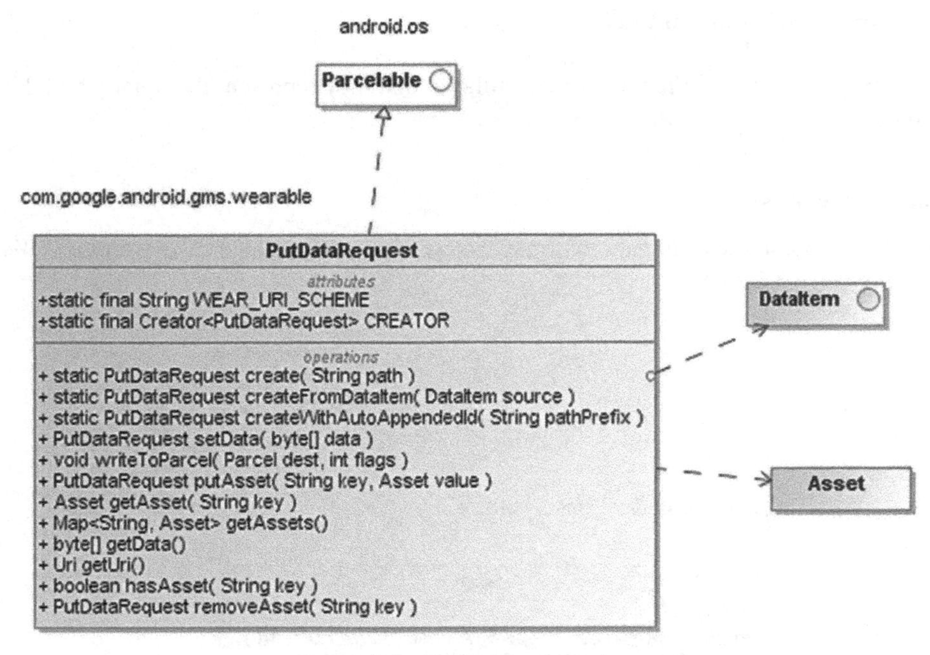

Figure 7-9 PutDataRequest class.

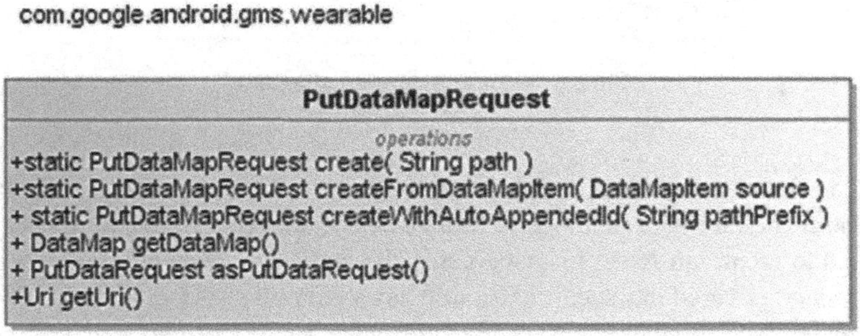

com.google.android.gms.wearable

PutDataMapRequest

operations
+static PutDataMapRequest create(String path)
+static PutDataMapRequest createFromDataMapItem(DataMapItem source)
+ static PutDataMapRequest createWithAutoAppendedId(String pathPrefix)
+ DataMap getDataMap()
+ PutDataRequest asPutDataRequest()
+Uri getUri()

Figure 7.10 PutDataMapRequest class diagram.

As shown in Figure 7.10, the *PutDataMapRequest* provides static methods to create its instance. Once instantiated, the *getDataMap()* method provides access to the encapsulated *DataMap*. Once the *DataMap* instance has been set with data as intended, the *asPutDataRequest()* method creates a *PutDataRequest* instance with the encapsulated data. After that, calling the *DataApi*'s *putDataItem()* method makes the request to add the data to the *Android Wear Network*. The *DataApi* does not have a put method that caters to the *PutDataMapRequest* directly.

There are far fewer methods in *PutDataMapRequest* as compared to *PutDataRequest*, because it's the *DataMap* operations that address the setting of the data.

7.6 Asset and DataItemAsset

Both the *Asset* and *DataItemAsset* are similar in that they represent the binary, blob kind of data.

7.6.1 Asset class

The *Asset* class represents data that may not yet have been added to the *Android Wear Network* (Figure 7-11A).

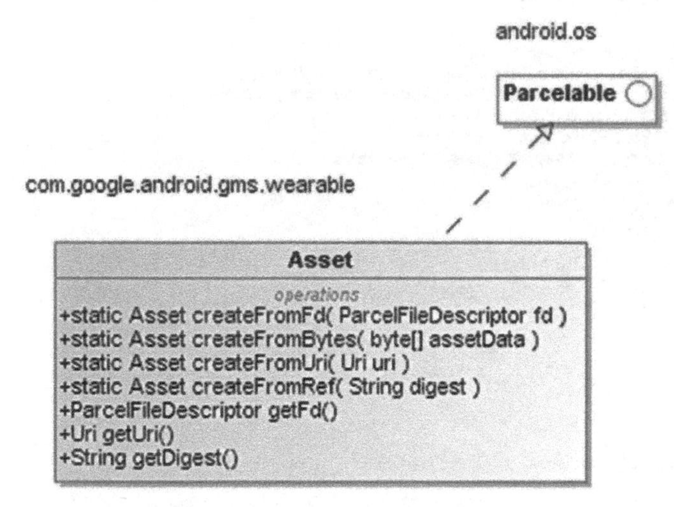

Figure 7-11A Asset class diagram.

The *Asset* class implements the *Parcelable* interface. The *Asset* instance can be created via its static create family of methods. A bitmap can be converted into a byte stream, which can be used to create an Asset instance using the static *createFromBytes()* method. The *PutDataRequest* covered in the earlier section has a method *putAsset()*, which can be used to populate the asset into the request. The *DataApi* has a *putDataItem()* method, which can be used to make the request to add data to the *Android Wear* Network. Alternatively, the *Asset* instance can be added to a *DataMap* instance via its *putAsset()* method, and the *PutDataMapRequest* can be used to create the *PutDataRequest* eventually.

The *getDigest()* method returns a digest—a one way hash of the content—which can be used to identify the *Asset* across devices on the *Android Wear Network.*

Assets can be received/extracted from data change events in the listener callbacks.

7.6.2 DataItemAsset interface

The *DataItemAsset* is a reference to an *Asset* after it has been added to the *Android Wear Network* as part of a *PutDataRequest* (Figure 7-11B).

The *getId()* method returns a unique identifier for the *Asset* in the *Android Wear Network.* The *getDataItemKey()* returns an identifier for the asset in the context of an existing *DataItem.*

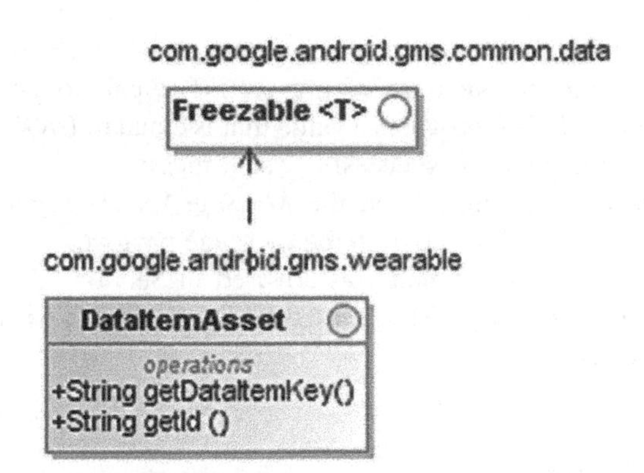

Figure 7-11B DataItemAsset interface diagram.

7.7 MessageApi

The *MessageApi* supports sending and receiving short, transient messages privately between instances of the same App that reside on different nodes in the *Android Wear Network*. A given message is private to the App that created it and only receivable on other instances of the App running on other nodes.

Messages are sent on a fire-and-forget basis—messages are delivered only to nodes that are currently connected to the *Android Wear Network*. As nodes join and leave the *Android Wear Network*, they will miss the messages that were sent during the time they had dropped off from the *Android Wear Network*. Therefore, a message pertaining to starting an Intent on the Wear device from the handheld, or a message pertaining to pausing the handheld device's media player, are examples where messages can be useful.

The *MessageApi* is suitable only for transient, short messages. On the other hand, the *DataApi* covered in the earlier section is appropriate for persistence and synchronization of long-lived data.

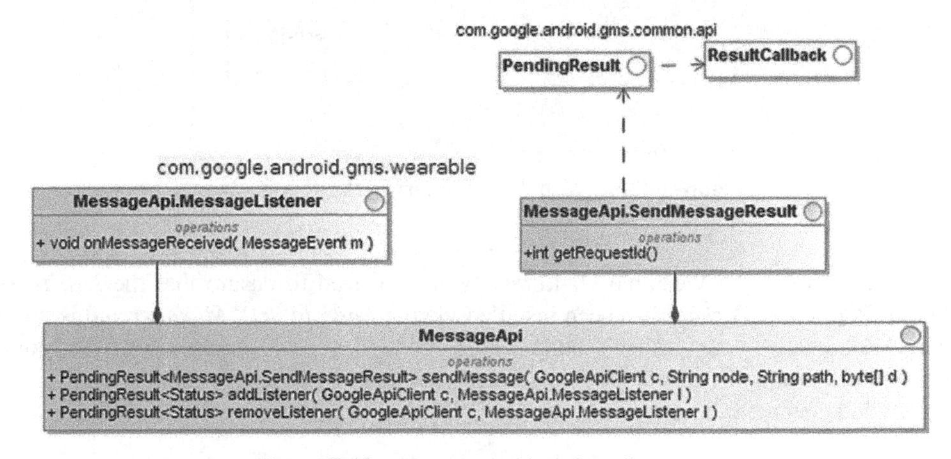

Figure 7-12 MessageApi interface diagram.

Figure 7-12 show the *MessageApi* interface and its nested interfaces *MessageApi.SendMessageResult* and *MessageApi.MessageListener*.

The *sendMessage()* method returns a *PendingResult* via which the asynchronous call back *onResult()* tells you the outcome of the call. The call to *getRequestId()* on the *MessageApi.SendMessageResult* provides a value that is equal to *UNKNOWN_REQUEST_ID* in case of failure or the ID of the successfully sent message.

The *onMessageReceived()* callback on the *MessageApi.MessageListener* is used for listening for *MessageEvents*, which contain the message payload.

The *WearableListenerService*, which was covered in Section 7.3.1.2, implements the *MessageApi* as well as the *DataApi*. The *WearableListenerService* is suitable for listening for events when in the background.

7.8 Wearable UI Library

Android Wear represents a unique form factor, and therefore, Wear Apps are significantly different from those for the handheld Android devices, both in terms of overall application design and in terms of user interface (UI).

The Wearable UI library (Figure 7-13A) is a support package that provides UI and supporting component exclusively for *Android Wear* Apps. It includes classes such as *WearableListView*, WatchViewStub, *CircledImageView*, *CardFragment*, and so on.

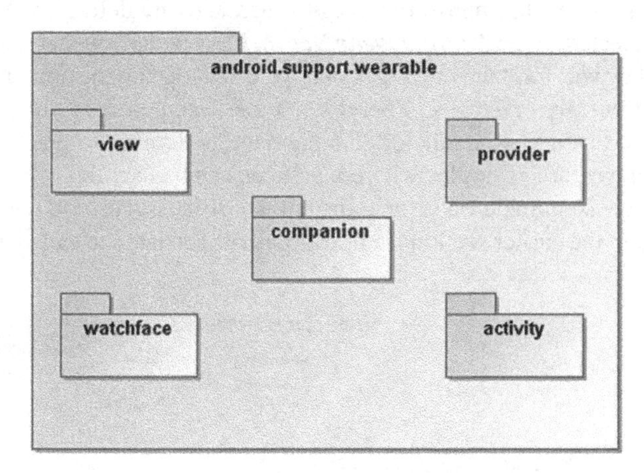

Figure 7-13A Wearable UI library package overview.

In order to use the Wearable UI library, you will need to ensure that the *Extras → Google Repository Package* has been installed via the *Android SDK Manager* and is up to date. Additionally, your project's build.gradle file will need to declare compile "com.google.android.support:wearable:+" in the dependencies section.

Detailed documentation on the Wearable UI can be found at:

https://developer.android.com/shareables/training/wearable-support-docs.zip
https://developer.android.com/training/wearables/ui/layouts.html

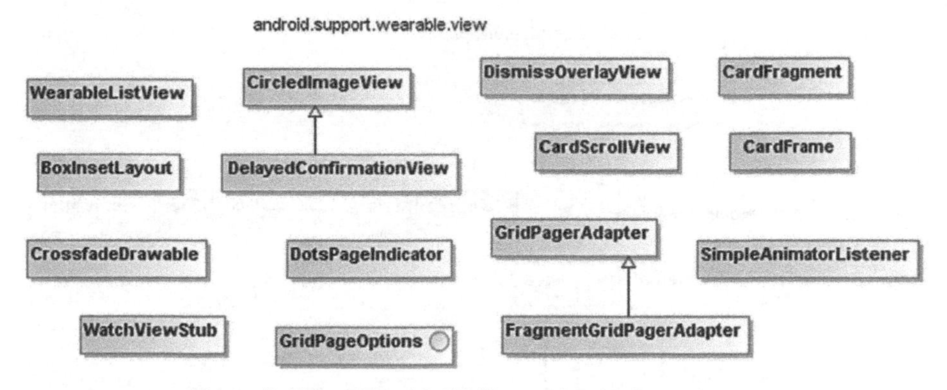

Figure 7-13B Wearable UI library view package classes.

Figure 7-13B shows the *WearableListView*, the *CircledImageView*, and several other classes in the *view* sub-package.

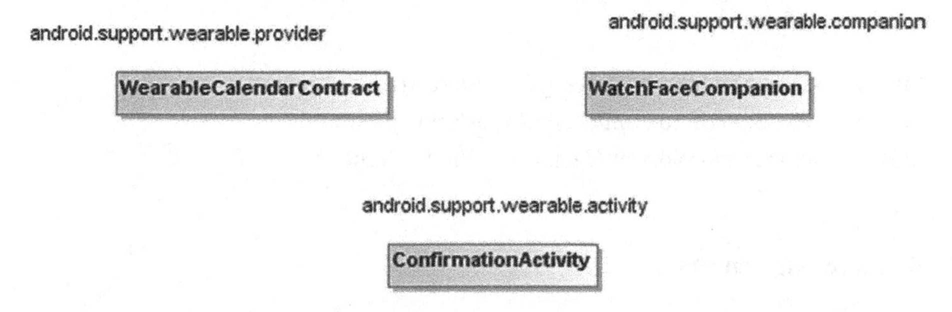

Figure 7-13C Wearable UI library provider, companion, and activity sub-packages.

Figure 7-13C shows the *WearableCalendarContract*, *WatchFaceCompanion*, and *ConfirmationActivity* classes in the *provider*, *companion*, and *activity* sub-packages, respectively.

Figure 7-13D shows the *WatchFaceService* and the *WatchFaceStyle* in the *watchface* sub-package, respectively.

7.9 Wear Interaction Design

Android Wear Apps are meant to be aware of the user's current context, physical location, activity, time of the day, and so on and provide relevant and timely information to the user. *Android Wear* is meant to act like a user's personal assistant. Thus, Wear Apps are meant to launch automatically and insert relevant cards into the context stream, provide simple, "glanceable" information, and require minimal interaction. The Wear design principles focus on not stopping the user and minimize vibration and interruption.

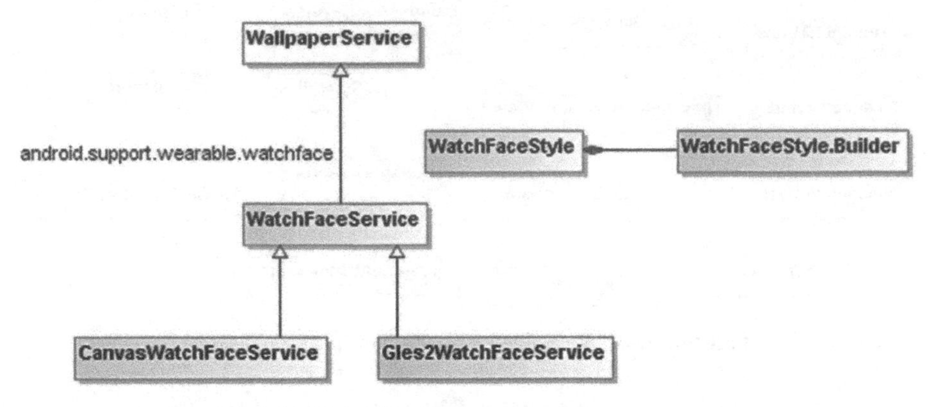

Figure 7-13D Wearable UI library watchface package classes.

The recommended creative vision and design principles for *Android Wear* Apps are covered elegantly at

https://developer.android.com/design/wear/creative-vision.html
https://developer.android.com/design/wear/principles.html
https://developer.android.com/design/wear/index.html

7.10 Accessing Sensors

Some *Android Wear* devices have built in GPS sensors, thereby providing the Wear device with independent location detection capabilities. Location awareness from within the Wear App can be useful in various contexts including but not limited to fitness. Wear Apps have direct access to the hardware sensors as well as user physical activities and the services that run on Wear device. As with any Android device, the PackageManager class provides the ability to query whether particular features and sensors are available on the device using the *hasSystemFeature()* method call.

In case the *Wear* device does not have a built in GPS and happens not to have an intact tethered connection to a handheld device (with location enabled), when it makes a call to access the location, it will encounter an exception and will need to hand that gracefully.

Since Wear devices are paired and tethered to handheld Android devices, they can depend on the handheld devices for various nuggets of information as needed. As long as the tethered connection is intact, a call to access the location on the Wear device from within the Wear App results in the Android OS determining and providing such information in the most optimal and power-efficient manner.

7.11 Production Wear Apps

With regard to the release and distribution of your Wear App, Wear devices do not offer the Google Play Store to the user. Therefore, users cannot browse, list, and install Wear Apps via their Wear devices directly from the Play Store.

In order to release an *Android Wear* apk to the Google Play Store, it will need to be packaged and embedded within the companion handheld App, which will need to be published to the Google Play Store. The *res/raw* directory of the handheld App project will need to contain the Wear App's Apk. Android Studio takes care of this for you when you create an App and target it to the handheld and *Android Wear* platforms simultaneously.

Wear Apps need to be released to the public via their "companion" handheld Apps published via the Google Play Store. Therefore, it is necessary to build and distribute a "companion" handheld App—with the Wear apk embedded within it. The embedded Wear App will be installed to the paired and tethered Wear device via the companion handheld App. This handheld App can also be useful for supporting the Wear App for accessing the network and providing it other services and functionality such as GPS and location info.

References and Further Reading

https://developer.android.com/google/auth/api-client.html

https://developer.android.com/google/play services/setup.html

https://developer.android.com/reference/com/google/android/gms/wearable/package-summary.html

https://developer.android.com/training/building-wearables.html

https://developer.android.com/training/wearables/ui/layouts.html

https://developer.android.com/shareables/training/wearable-support-docs.zip

https://developer.android.com/design/wear/creative-vision.html

Part IV Google Fit Platform and SDK

This section covers the *Google Fit* platform and API. The user's ecosystem may include a multitude of devices, sensors, and Fit applications from a diversity of vendors. In such a backdrop, Google Fit provides a user centric data repository and common data types in the interest of interoperability and collaboration. Google Fit applications are aimed at a very specific segment and are bound by very specific terms and conditions. For one, *Google Fit* applications are not intended to be "medical" applications. The *Google Fit* platform is not intended to provide biometric functions. There are several developer conditions and obligations regarding the use of the *Google Fit* API. This section covers the hands on steps of connecting a Bluetooth LE fitness sensor and connecting to it from the Google Fit platform.

Wearable Android™: Android Wear & Google Fit App Development, First Edition. Sanjay M. Mishra.
© 2015 John Wiley & Sons, Inc. Published 2015 by John Wiley & Sons, Inc.

Chapter 8 Google Fit Platform

8.1 Google Fit Platform Overview

The base Android platform SDK has had since some time the capabilities for accessing built-in sensors and connecting with peripherals including fitness sensors via mechanisms including Bluetooth, Bluetooth LE, and more; these were covered in Chapter 5. Fitness-tracking Apps for Android have preceded the release of the *Google Fit* platform. The *Google Fit* platform takes fitness-tracking Apps to a more sophisticated level—it addresses the domain from end to end for application developers.

Google Fit is a platform and open ecosystem for fitness and wellness tracking. A multitude of sports & apparel companies, hardware vendors, and prominent App developers have partnered with Google to collaborate and participate in the *Google Fit* platform. At the time of writing, the *Google Fit* partners include companies such as Nike, HTC, LG, Withings, Motorola, Noom, Polar, Runstatic, and RunKeeper. *Google Fit* helps developers in building sophisticated fitness Apps while giving users complete control over their data.

Google Fit is user centric and hardware and application vendor independent—such that fitness readings from diverse sensors can be acquired, categorized, stored, shared, and analyzed in a collaborative and secure manner. The consumer has complete control over the sharing of their data; they can also delete the data at any time. The cloud-based *Google Fit* API supports Android via the Fit API while also supporting Web and iOS clients via a REST API.

Wearable Android™: Android Wear & Google Fit App Development, First Edition. Sanjay M. Mishra.
© 2015 John Wiley & Sons, Inc. Published 2015 by John Wiley & Sons, Inc.

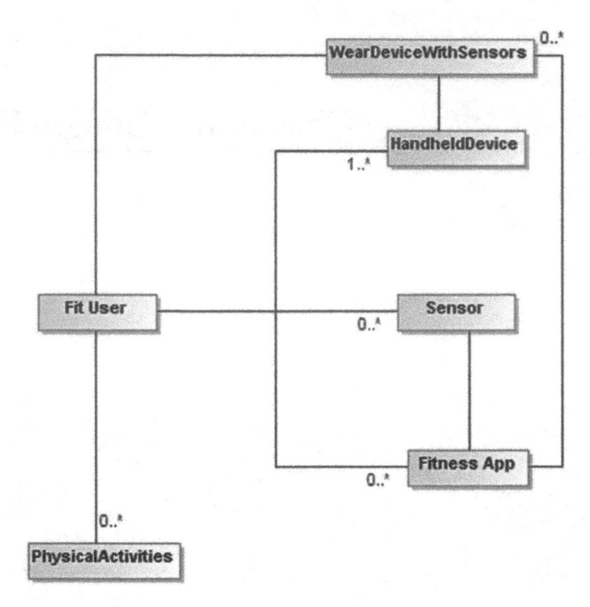

Figure 8-1 Users, devices, sensors, and fitness Apps.

Figure 8-1 attempts to model the relationships and correlations between Users, handheld devices, sensors, and Apps. A given user can have multiple devices on their person while engaging in some real-world physical activities. The devices on their person may have hardware sensors as well as Fit Apps installed on them. The real-world physical activities that the user engages in can be inferred by the hardware sensors and software stacks. In such a backdrop, of a multitude of devices and Apps from various vendors, it is important to have a common data schema in the interest of interoperability and collaboration. It is also important to organize fitness data in a user-centric data repository. Not shown in the diagram—to keep it simple initially—is the sensor data associated with the user that can be stored on the cloud-based Fit store.

8.2 Google Fit Core Concepts

The *Google Fit* API provides applications with the interfaces to access sensor data, record and store such data, and access history data—all subject to user permission and control. *Google Fit* provides important features such as standardized **Fit Data Types**, **Sensor** Data Access and **Recording** support, cloud **Fit data Store** (storage), and **History** data access.

8.3 Fit Data Types

The nature of fitness sensor readings brings up the question of the units of measure: are the various quantities to be measured in meters or feet, miles or kilometers, counts per minute or hour, Lbs or Kilograms, … ? *Google Fit* **Data Types** are standardized data types that help establish a common data schema and units of measure. Fit Data types provide an off-the-shelf schema that helps App developers jump start App development and boosts interworkability between Apps and devices from diverse providers, vendors, and App publishers. A given user may have several fitness sensors from various manufacturers on their person. The user may also use a multitude of fitness Apps—perhaps to evaluate them or because particular Apps have a better user experience or better support for particular fitness activities that they engage in, and so on. Fit Data types standardize on the units of measure for various quantities such as heart rate in beats per minute, weight in kilograms, etc. Fit Types happen to be aligned with metric system, which means that units such as meters and kilograms are used for storage.

Much like the fundamental data types in programming languages identify and classify the basic types of data such as *int*, *long*, *char*, *byte*, *float*, etc. and define their meanings, ranges of values that they can hold, and the operations that can be performed on them, so also the Fit Data types help classify and define relevant fundamental types of data in the fitness domain, such as height, weight, heartbeat, etc, their associated units of measure, and so on. Fit data types make it possible for different Apps to interoperate seamlessly using these standard types.

Thus, *Google Fit* data types help establish a common data schema that's usable across fitness device models and fitness Apps. *Google Fit* Data types define a class model that addresses fundamental fitness types such as height, weight, heart rate, speed, etc.

8.4 Fit Data Store (Storage)

Google Fit **Store** (Storage) is a cloud service for Fit data storage hosted on Google's infrastructure. The Fit Store provides secure data storage and access capabilities for Mobile Apps and Browser clients alike.

Fit Apps can access data stored in the Fit Store by other Apps, that is, data on the Fit Store can be shared between different Apps—subject to the user's approval and permission. User consent is necessary in order for any Fit App to read or write fitness data. Access to data is governed by Authorization and Permissions—users can exert fined grained control over what particular types of fit data an App can access and whether in read or write mode. The authorization and permissions are implemented using scopes which leverage the OAuth 2.0 Scope to regulate access from particular Apps to the Fit Store.

8.5 Sensors

The Sensor API helps Apps access raw sensor data from sensors available in Android handheld devices, companion devices such as Android Wear smart watches, and peripheral fitness sensor devices.

8.6 Permissions, User Consent

Google Fit Apps require the user's consent before they become operational and commence to read and save *Google Fit* data. There are different categories of data, and *Google Fit* provides users with granular control and consent mechanisms. Thus, users have the ability to allow an App to access fitness data of particular categories in read or write mode. *Google Fit* provides a *Permissions* model and ***Scope*** of data access, so that it is transparent to the user what data—specific sensor readings and history records—a particular fitness App is accessing.

8.6.1 Permission Groups, Fitness Scopes

Google Fit defines OAuth scopes, which map to permission groups: activity, body, and location. Each of these permission groups maps to specific data types. When Apps specify a scope that they work with, *Google Fit* in turn makes the request of the corresponding permission to the user. Only after the user's consent is received, the *Fit* App is able to do its work in compliance with the received consent.

Google Fit's permission mechanism is handled at runtime and the first time that the Fit App is run. This is slightly different but analogous in concept to the standard Android Permission mechanism—when an App declares particular permissions such as "*android.permission.RECORD_AUDIO*" or the more commonly used "*android.permission.INTERNET*" in the application's *AndroidManifest.xml* file, these permissions are exposed to the user via the Google Play Store's catalog prior to installation; furthermore, after installation, the same permissions—and nothing more—are granted to the App at run time in the secure, sandboxed environment on the user's Android device.

8.6.1.1 Activity Scope The Activity scope pertains primarily to the data associated with the type of the user's physical activity or sport such as basketball, tennis, swimming, walking, running, cycling, skiing, yoga, gardening, sleeping, and so on. It also includes calories consumed and such information associated with the activity and its duration. More information can be found at:

https://developers.google.com/fit/android/authorization#fitness_scopes.

The *Google Fit* Activity represents physical Activity and is unrelated to the user interface Activity. It's easy to discriminate between the two based on the context of the conversation.

8.6.1.2 Body Scope The Body scope pertains to information about the user's body metrics such as heart rate, weight, height, and so on.

8.6.1.3 Location Scope The Location scope pertains to the user's physical location, speed, and so on.

8.7 Google Fit: Developer Responsibilities

App developers who build Apps for *Google Fit* are advised to adhere to certain core values:

Transparency
: Make the user aware of what data is being collected and why.

Fitness and wellness purposes only
: Do not use the data for purposes other than fitness (such as biometric, medical, commercial, advertising, etc).

Adherence to user requests
: Ensure that user requests such as deletion of their data are honored.

Adherence to *Google Fit* Developer's Terms and Conditions and Branding Guidelines
: which covers all of the above and more…

8.7.1 Developer Terms and Conditions

The *Google Fit* Terms and Conditions are available at https://developers.google.com/fit/terms.

8.7.2 Developer Branding Guidelines

The *Google Fit* Branding Guidelines are available at https://developers.google.com/fit/branding.

8.8 Procuring Sensor Peripherals

Your *Android Wear* device as well as your handheld Android device likely has several sensors related to fitness including step counters, heart rate monitors, and more. In case you have a Bluetooth LE sensor for fitness tracking, that would be useful for your Fit development experiments. You may consider purchasing any Bluetooth LE fitness sensor that supports a standard Bluetooth Low Energy GATT profile. In case you have a Bluetooth classic fitness device, that would be useful for writing a software sensor.

8.9 Hello Fit: hands-on example

Now that we have somewhat familiarized ourselves with the *Google Fit* platform and its ground rules, we are ready to write our simple "Hello Fit" App. The corresponding project *hellofit* is available in its entirety, as part of sample code for this book. This project was created manually and uses the classic, *ant*-based project tree, while the rest of the *Fit* samples that make up the sample code for this chapter use the *gradle*-based project tree. As an Android developer, you will likely inherit and need to maintain existing projects or leverage prominent library projects that happen to be aligned with the classic project tree. Therefore, it is virtually essential at this time to be proficient in both these project structures. (*Android Studio* is my default choice of IDE for Android development and does a great job of creating a new project and managing the dependencies between "modules"—modules is a recently introduced term. There are various types of modules such as Android Application, Library, Test, and so on. Most of this book was written before the term modules became a formally defined term.)

When creating a *Google Fit* App, there are several setup steps and prerequisites including using the application's signing keystore's SHA1 certificate fingerprint and the application's package name to enable the Fit API via the *Google Developer Console*. Also, Fit projects have a compile time dependency on the *Google Play Services* library project.

The Hello Fit App accesses fitness data from available sensors using the *SensorsApi*. It registers interest in data points of data types heart rate bpm, speed, and cumulative steps. You may change these data types per your interests and to match the sensors that you have available. The details of the *Google Fit* API are covered in the next chapter.

8.9.1 Google Play Services library project, dependency

All *Google Fit* projects depend on the *Google Play Services* library project, which is part of the *Extras* that are available in your Android SDK installation.

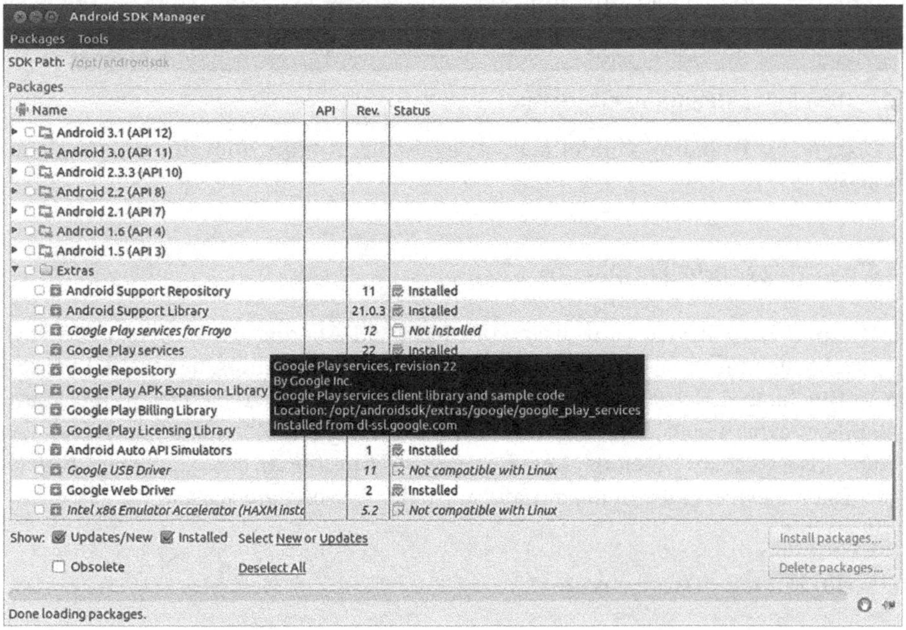

Figure 8-2A Extras: Google Play Services.

Figure 8-2A shows the *Extras* section in the Android SDK Manager which includes the *Google Play Services* library project with it's installation status indicator in my local development environment. You will need to ensure that *Google Play Services* has been installed and is up to date in your local development environment before proceeding further.

I copied over the *Google Play Services* library project into a *projects/external* area in my home directory, so as to avoid working within the Android SDK installation tree. The following snippet shows the steps that built the Google Play Services library project via the command line. The target of 17 is a relative target that happens to represent API level 21 (Android 5.0) in my local installation.

```
$ cp -r /opt/androidsdk/extras/google/google_play_services/libproject/google-play-services_lib/.
$ mv google-play-services_lib google-play-services
$ cd google-play-services/
$ android list targets
...
$ android update project --path . --target 17
$ cat projects.properties
        # Project target.
        target=Google Inc.:Google APIs:21
        android.library=true
$ ant clean debug
```

Figure 8-2B shows the setup of the play services library project. Any classic Android project tree can be enabled for building via *ant* by executing the android update project command.

```
~/projects/external/google-play-services $ android update project --path . --target 17
Updated project.properties
Updated local.properties
Updated file ./proguard-project.txt
It seems that there are sub-projects. If you want to update them
please use the --subprojects parameter.
~/projects/external/google-play-services $ cat project.properties | tail -2
target=Google Inc.:Google APIs:21
android.library=true
~/projects/external/google-play-services $ ant clean debug
```

Figure 8-2B Updating play services library project.

Figure 8-2C shows the successful building of the Google Play Services library project.

In case you are using a gradle-based build, you will need to declare the dependency on *Google Play Services* via an entry in the App's *gradle.properties* as shown in the snippet below:

```
dependencies {
  ...
  compile ' com.google.android.gsm:play-services:6.5.87'
}
```

You will need to update this version to the latest version, which at the time of writing happens to be 6.5.87.

```
-package:
     [echo] Library project: do not package apk...

-post-package:

-do-debug:
     [echo] Library project: do not create apk...
[propertyfile] Creating new property file: /home/sanjay/projects/external/google-play-services/bin/buil
d.prop
[propertyfile] Updating property file: /home/sanjay/projects/external/google-play-services/bin/build.pr
op
[propertyfile] Updating property file: /home/sanjay/projects/external/google-play-services/bin/build.pr
op
[propertyfile] Updating property file: /home/sanjay/projects/external/google-play-services/bin/build.pr
op

-post-build:

debug:

BUILD SUCCESSFUL
Total time: 3 seconds
~/projects/external/google-play-services $
```

Figure 8-2C Building play services library project.

8.9.2 Using the SHA1 fingerprint of the keystore

You will need to determine the SHA1 fingerprint of the certificate that your Fit App will be signed with. During development, Android applications are signed by the debug keystore by default. The debug keystore is located at ~/.android/debug.keystore. The keystore password is "android":

```
$ keytool -list -keystore  ~/.android/debug.keystore
        Enter keystore password:
        Keystore type: JKS
        Keystore provider: SUN
        Your keystore contains 1 entry
        androiddebugkey, Sep 14, 2014, PrivateKeyEntry,
Certificate fingerprint (SHA1):
9E:EB:6B:61:31:62:D7:80:5E:DE:0D:86:2D:B0:4F:4F:F9:CF:20:D9
```

You may also create a keystore for your Fit App, suitable for using for a release version; the snippet below shows the steps that this entails:

```
$ keytool -genkey -v -keystore wearbookfit.keystore -alias wearbookfitCert -keyalg
RSA -keysize 3072 -validity 50000
...
$ keytool -list -v -keystore wearbookfit.keystore
...
```

In either case, you must be sure to use the SHA1 fingerprint of the keystore that you will be using to sign your App.

Figures 8-3A and 8-3B show the commands I used to create the keystore, which I will use for signing the Hello Fit App.

```
~/projects/hellofit $ keytool -genkey -v -keystore wearbookfit.keystore -alias w
earbookfitCert  -keyalg RSA -keysize 3072 -validity 50000
Enter keystore password:
Re-enter new password:
What is your first and last name?
  [Unknown]:  Sanjay Mishra
What is the name of your organizational unit?
  [Unknown]:  self
What is the name of your organization?
  [Unknown]:  self
What is the name of your City or Locality?
  [Unknown]:  Los Gatos
What is the name of your State or Province?
  [Unknown]:  CA
What is the two-letter country code for this unit?
  [Unknown]:  US
Is CN=Sanjay Mishra, OU=self, O=self, L=Los Gatos, ST=CA, C=US correct?
  [no]:  yes
```

Figure 8-3A Keytool—creating a keystore.

```
~/projects/hellofit $ keytool -genkey -v -keystore wearbookfit.keystore -alias w
earbookfitCert  -keyalg RSA -keysize 3072 -validity 50000
Enter keystore password:
Re-enter new password:
What is your first and last name?
  [Unknown]:  Sanjay Mishra
What is the name of your organizational unit?
  [Unknown]:  self
What is the name of your organization?
  [Unknown]:  self
What is the name of your City or Locality?
  [Unknown]:  Los Gatos
What is the name of your State or Province?
  [Unknown]:  CA
What is the two-letter country code for this unit?
  [Unknown]:  US
Is CN=Sanjay Mishra, OU=self, O=self, L=Los Gatos, ST=CA, C=US correct?
  [no]:  yes

Generating 3,072 bit RSA key pair and self-signed certificate (SHA256withRSA) wi
th a validity of 50,000 days
        for: CN=Sanjay Mishra, OU=self, O=self, L=Los Gatos, ST=CA, C=US
Enter key password for <wearbookfitCert>
        (RETURN if same as keystore password):
```

Figure 8-3B Keytool—creating a keystore, continued.

Figure 8-3C shows the command for listing the SHA1 fingerprint.

```
~/projects/hellofit $ keytool -list -v -keystore wearbookfit.keystore
Enter keystore password:
```

Figure 8-3C Keytool listing.

8.9.3 Google Developer's Console Activating Fit API

Keeping the SHA1 fingerprint information readily available and after having ascertained the package name of the Hello Fit app (*io.wearbook.hellofit*), I proceeded to sign into the Google Developer Console at https://console.developers.google.com.

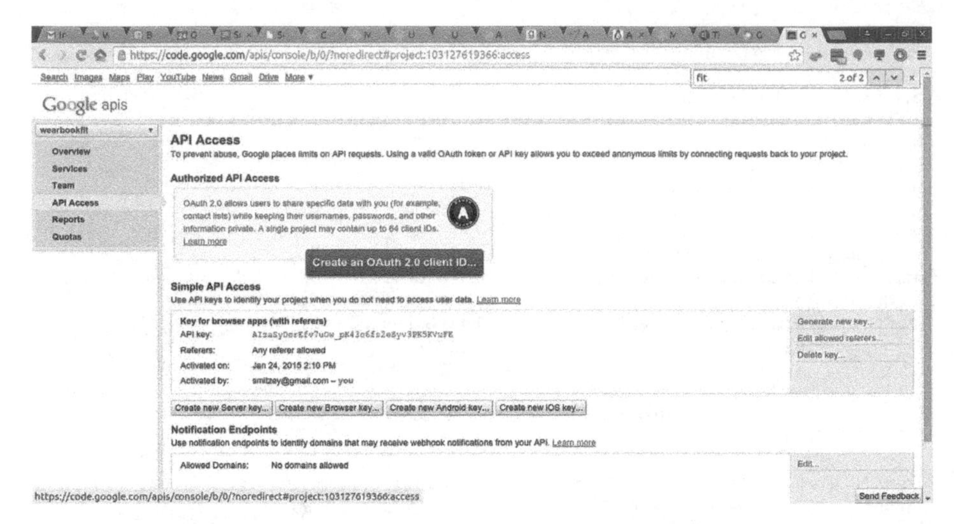

Figure 8-4A Google Developer Console, wearbookfit project.

Figure 8-4A shows the wearbookfit project that I created. I chose Android as the application type and entered in the SHA1 fingerprint and package name as shown in Figure 8-4B.

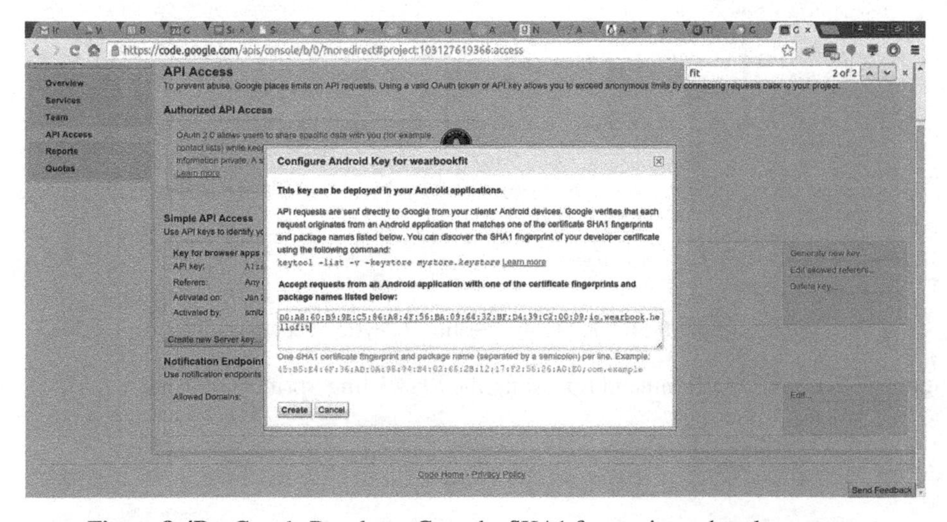

Figure 8-4B Google Developer Console, SHA1 fingerprint and package name.

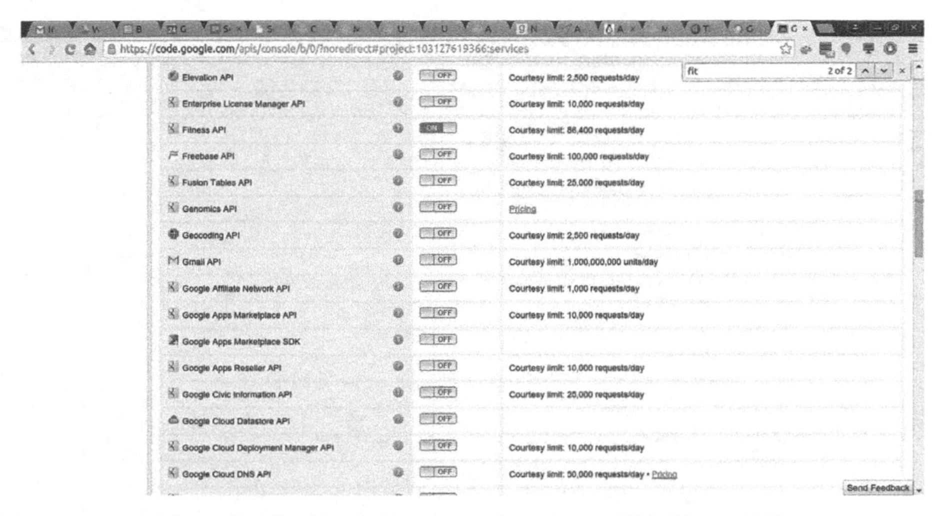

Figure 8-4C Google Developer Console, enabling Fitness API.

Figure 8-4C shows the enabling of the Fitness API.

Figure 8-4D shows terms and *Google Fit* terms of service that will need to be accepted before proceeding further.

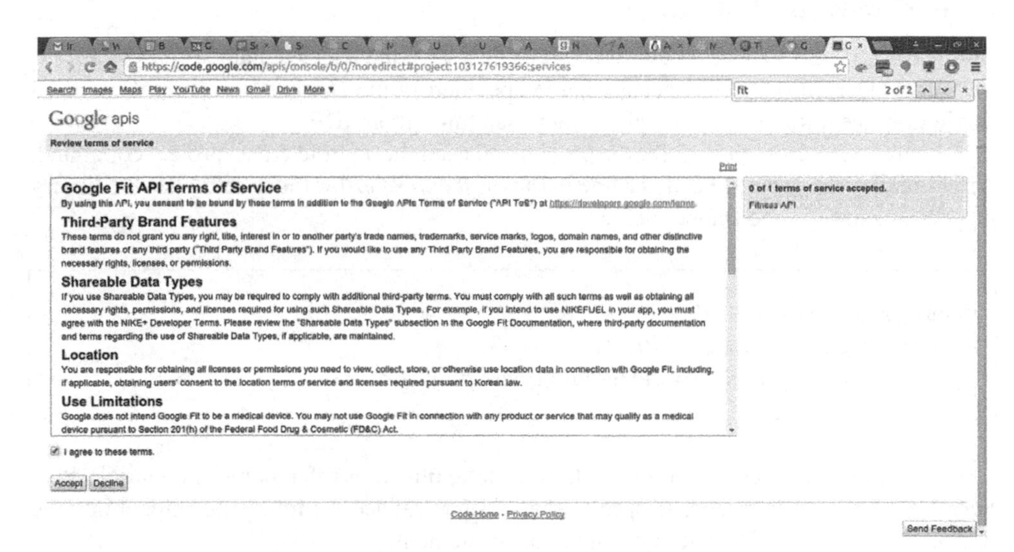

Figure 8-4D Google Developer Console, Fitness API terms.

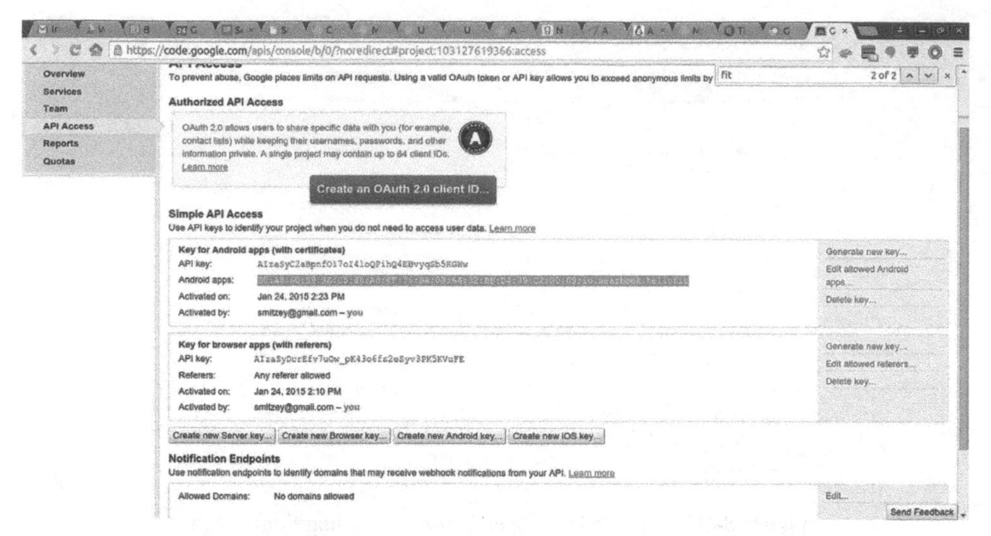

Figure 8-4E Google Developer Console, Fitness API setup complete.

Figure 8-4E shows the Console with the Android client now authorized and setup for *Google Fit*.

8.9.4 Creating the Android App

My next step was to create a new Android project with the package name of *io.wearbook.hellofit*, coincident with the package name specified in the *Google Developer Console*. I carried out the initial steps manually rather than through an IDE.

I created a directory *hellofit* for the project and used the android create project command. After creating the project, I added a line in the *project.properties* file to reference the library project as shown in the snippet below:

```
$ cat project.properties
target=Google Inc.:Google APIs:21
android.library.reference.1=../external/google-play-services
```

Google Play Services are versioned, and the application referencing the Google Play Services library project needs to specify the version. I added the following snippet into the *AndroidManifest.xml* file within the application element:

```
<meta-data android:name="com.google.android.gms.version"
    android:value="@integer/google_play_services_version" />
```

I copied the *version.xml* file from *res/values* in the Google Play Services library project into res/values within the hellofit project. The contents of *version.xml* is shown in the snippet below:

```
<?xml version="1.0" encoding="utf-8"?>
<resources>
  <integer name="google_play_services_version">6587000</integer>
</resources>
```

The following snippet shows the contents of *ant.properties*, which contains the references to the keystore that I will be using for this project:

```
~/projects/hellofit $ cat ant.properties
key.store=./wearbookfit.keystore
key.alias=wearbookfitCert
key.store.password=****
key.alias.password=****
```

The following snippet represents the steps that I executed to create the project:

```
$ mkdir hellofit
$ cd hellofit/
$ android list targets
$ android create project --path . --name hellofit --target 17 --package
io.wearbook.hellofit --activity HelloFitActivity
```

After making the changes shown earlier, the project built successfully:

```
$ ant clean release install
```

The next step was to edit the *HelloFitActivity* to access the Fitness API via the *GoogleApiClient*. The following snippet shows the initialization of the *GoogleApiClient*:

```
private void initGoogleApiClient() {
   this.googleApiClient = new GoogleApiClient.Builder(this)
         .addApi(Fitness.API)
         .addScope(new Scope(Scopes.FITNESS_BODY_READ))
         .addScope(new Scope(Scopes.FITNESS_LOCATION_READ))
         .addScope(new Scope(Scopes.FITNESS_ACTIVITY_READ))
         .addConnectionCallbacks(
               new GoogleApiClient.ConnectionCallbacks() {

               ......
```

Once upon connecting the *GoogleApiClient*, I scanned for Bluetooth LE devices as shown below:

```
PendingResult<Status> pendingResult = Fitness.BleApi.startBleScan(
                    googleApiClient,
                    new StartBleScanRequest.Builder()
                    .setDataTypes(DataType.TYPE_HEART_RATE_BPM)
                    .setBleScanCallback(bleScanCallback)
                    .build());
```

The *BleScanCallback* is an abstract class, which has methods *onDeviceFound* and *onScanStopped*. The following snipped shows my concrete implementation class that extends *BleScanCallback*:

```
class MyBleScanCallbackAndHelper extends BleScanCallback {

 @Override
 public void onDeviceFound(BleDevice device) {
  Log.d ( TAG, "MyBleScanCallback:onDeviceFound found device=" + device ) ;

  bleDeviceFound = device ; // add to list of devices TODO

  PendingResult<Status> pendingResult =
          Fitness.BleApi.claimBleDevice(googleApiClient, device);
 addContentToView("MyBleScanCallback:onDeviceFound foundBleDevice=" + device);
  Status status =     pendingResult.await() ;
  // resolve unsuccessful status here ....
  Log.d (TAG, "MyBleScanCallback:onDeviceFound status=" + status ) ;
  // error checks to be done here ...
  registerDataSourceListener(DataType.TYPE_HEART_RATE_BPM);

 }
 @Override
 public void onScanStopped() {
  Log.d ( TAG, "MyBleScanCallback:onDeviceFound found device=" + device ) ;
 }

 void releaseDevice ( BleDevice device ) {
  Fitness.BleApi.unclaimBleDevice(googleApiClient, device);
 }
```

I also registered several data types with an *OnDataPointListener.*

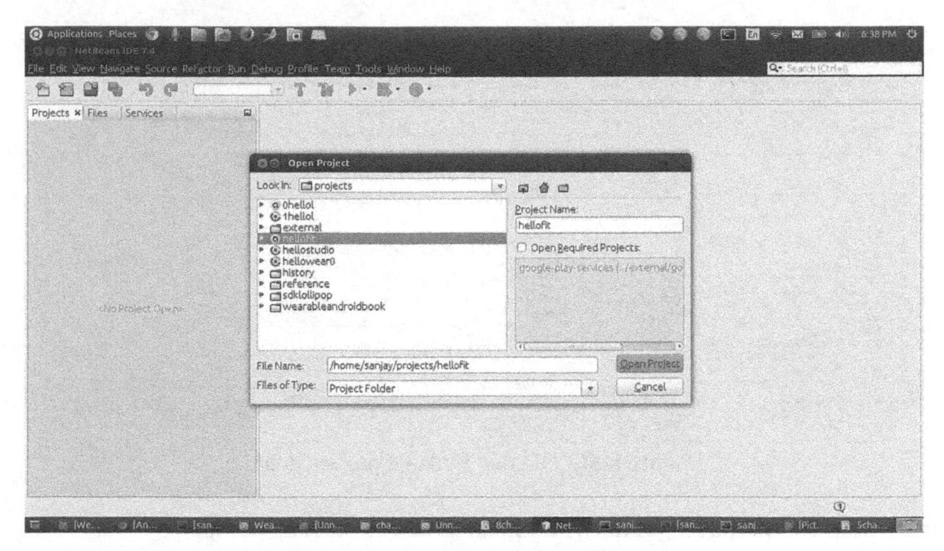

Figure 8-5A Editing Hello Fit project.

Figure 8-5A shows the opening of the hellofit project using NetBeans, which recognized the Android classic project tree and hence displayed the “a” icon seen in the figure.

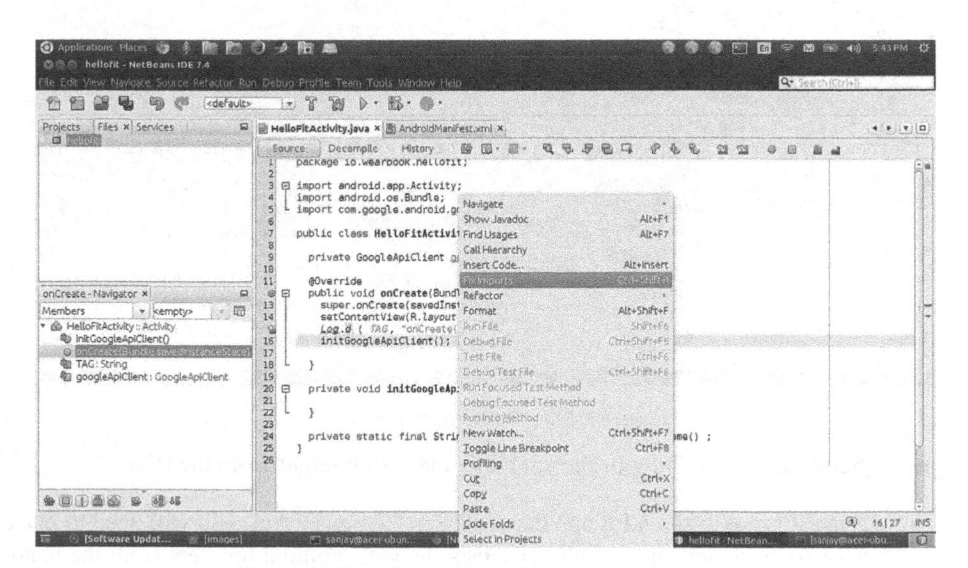

Figure 8-5B Editing Hello Fit project, continued.

Figure 8-5B shows the *HelloFitActivity* early in the editing process.

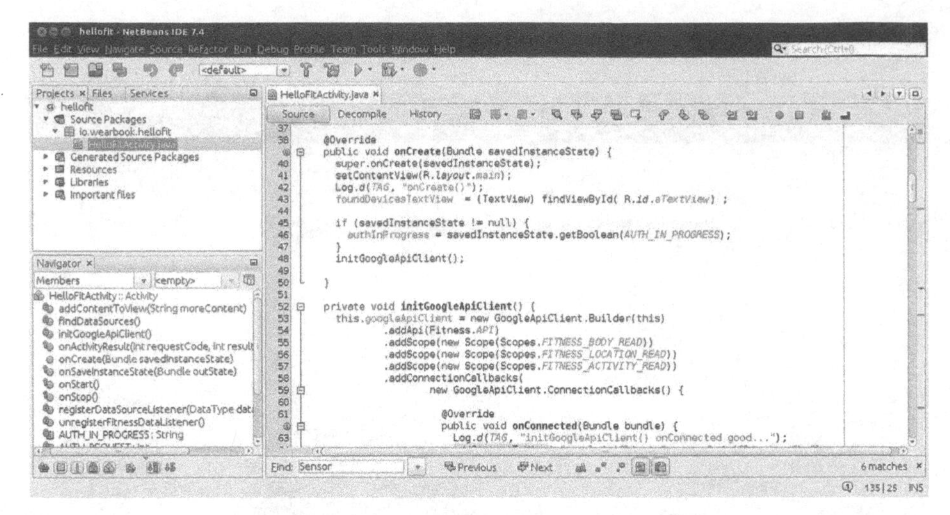

Figure 8-5C Editing Hello Fit project, done.

Figure 8-5C shows the HelloFitActivity in an advanced stage of editing.

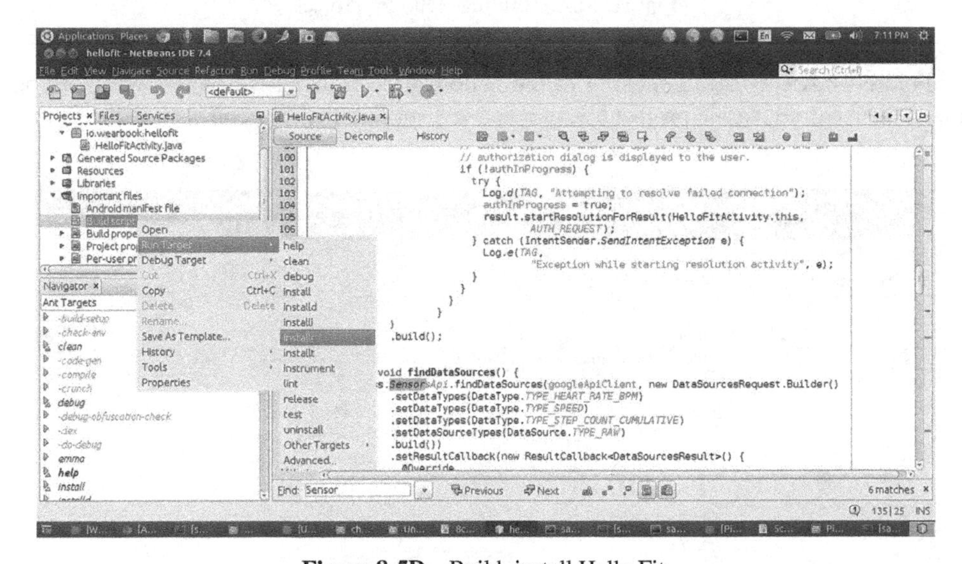

Figure 8-5D Build, install Hello Fit.

Figure 8-5D shows the running of the ant build and install target from the IDE.

Before you attempt to run a Fit app that accesses the SensorApi, you will need to place your Android Wear device and peripheral sensor devices within a few feet from the handheld device that has Fit app installed. It does not hurt to pair the sensor/peripheral device with your handheld device.

After installing the Hello Fit App on my Android handheld device, I strapped on my Zephyr Heart Rate Monitor and wore my Android Wear smart watch device since it has a few sensors.

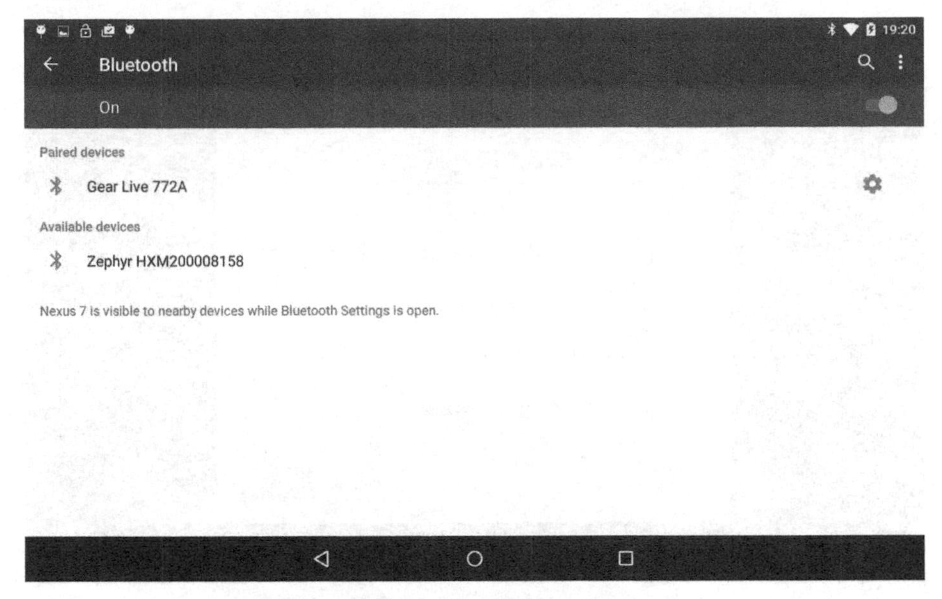

Figure 8-5E Zephyr Heart Rate Monitor.

Figure 8-5E shows my Zephyr heart rate monitor visible to my handheld Android device. I subsequently paired the heart rate monitor (not shown in the figure). After that, I started the Hello Fit App.

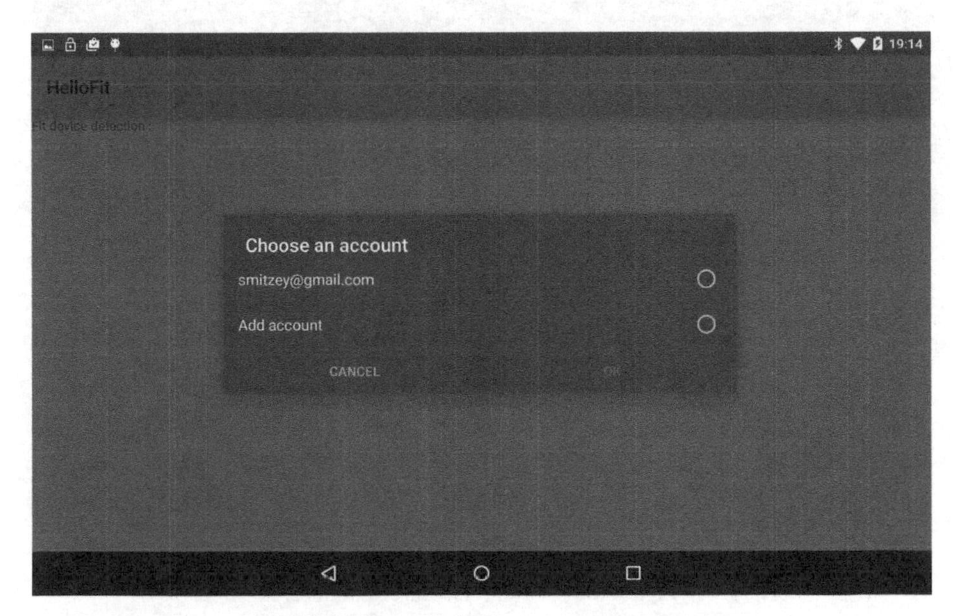

Figure 8-6A Hello Fit—choose account.

Figure 8-6A shows the Hello Fit app requesting an account to be selected or added.

Figure 8-6B Hello Fit—permissions.

Figure 8-6B shows the user the scope of the access to fitness data. After I accepted the application's access to the specified scopes, the activity *HelloFitActivity* was displayed.

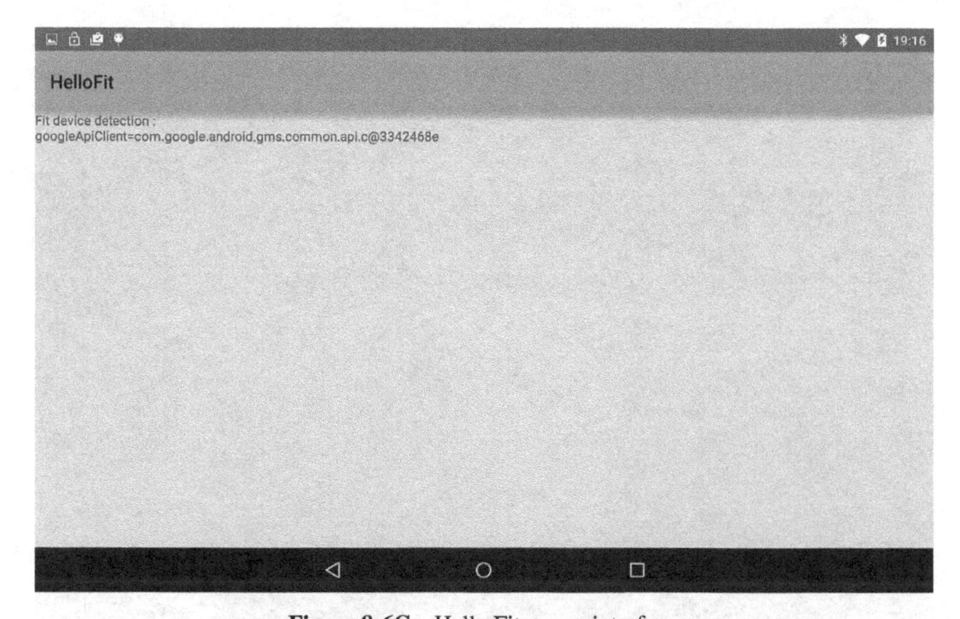

Figure 8-6C Hello Fit—user interface.

Figure 8-6C shows the *HelloFitActivity* with the steps data point. After I walked around a few steps, I found the step counts incrementing.

Figures 8-6D, 8-6E, and 8-6F show the *HelloFitActivity* with several increments of the steps data points. *Android Wear* devices have a step counter and so do several models of handheld devices.

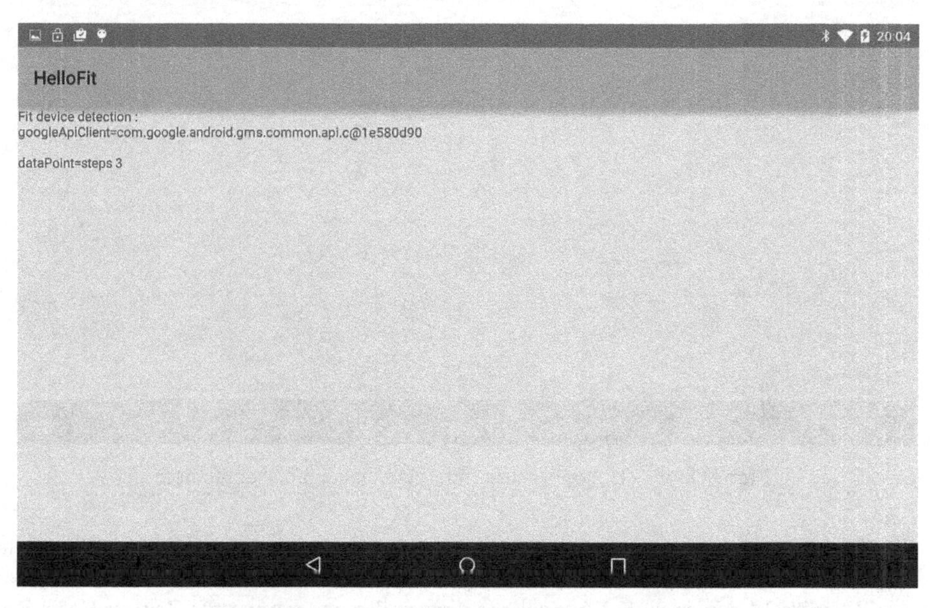

Figure 8-6D Hello Fit—user interface, continued.

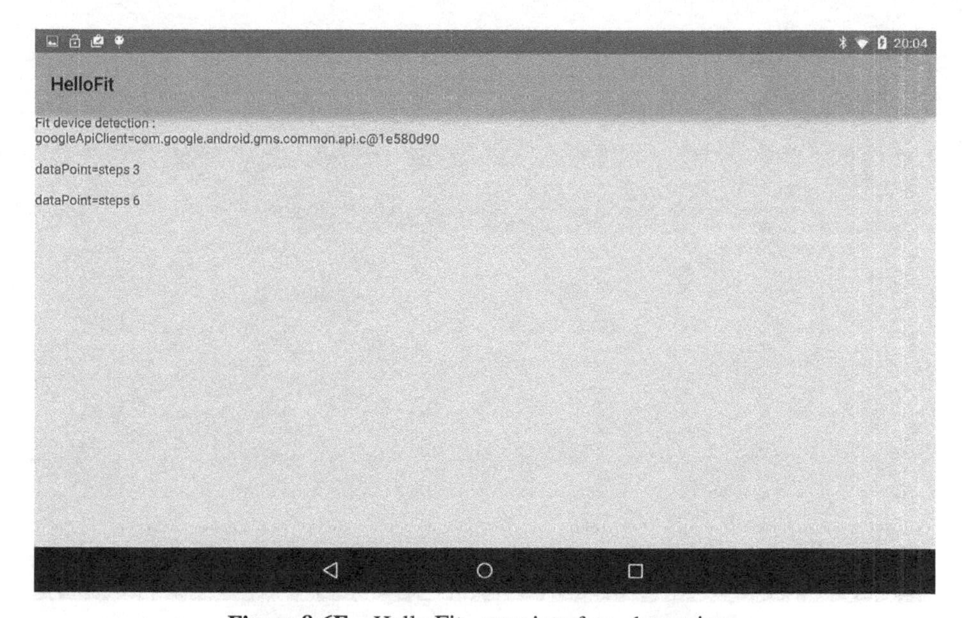

Figure 8-6E Hello Fit—user interface, data points.

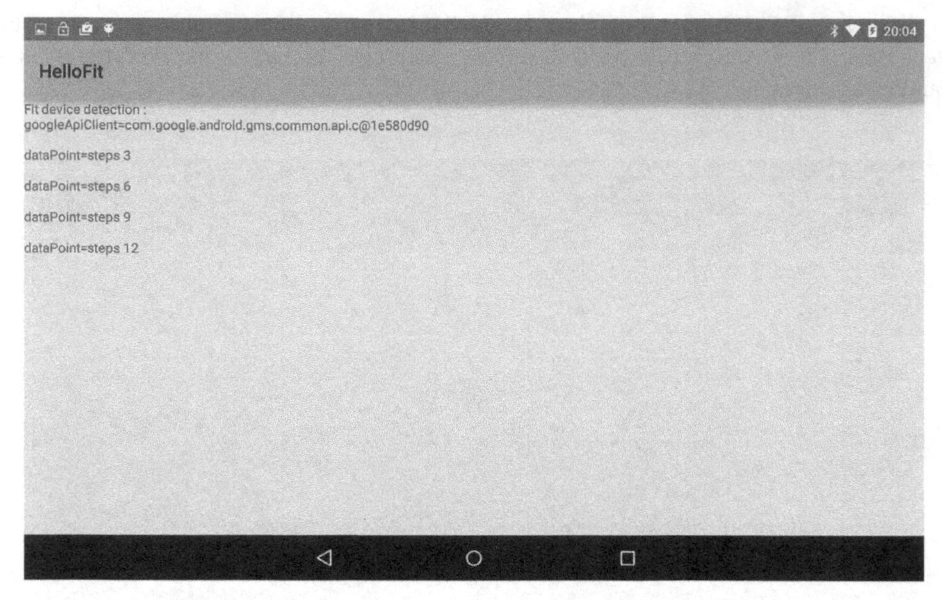

Figure 8-6F Hello Fit—user interface, data points, continued.

So far, I did not have success in detecting the heart rate monitor. I verified that I had paired my handheld device with the Zephyr Heart Rate Monitor. After rebooting my Android handheld device and also disconnecting and reconnecting the Zephyr Heart Rate Monitor's chest strap, I had some success. Subsequently, I found that the heart rate sensor was not detected by the App consistently.

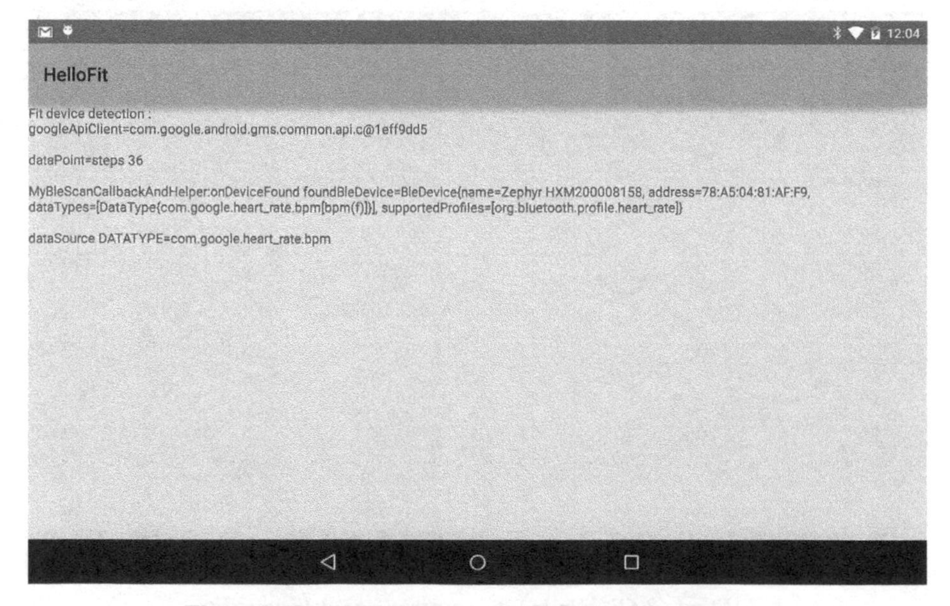

Figure 8-6G Hello Fit—user interface, heart rate data type.

Figure 8-6G shows the Zephyr Heart Rate Monitor detected by the Hello Fit app. The logs snippets show a few of the steps pertaining to the successful scanning and detection of the data source corresponding to the Zephyr Heart Rate Monitor device:

```
MyBleScanCallbackAndHelper:onDeviceFound found device=BleDevice{name=Zephyr
HXM200008158, address=78:A5:04:81:AF:F9,
dataTypes=[DataType{com.google.heart_rate.bpm[bpm(f)]}],
supportedProfiles=[org.bluetooth.profile.heart_rate]}
...

findDataSources onResult() DataSource{raw:Zephyr HXM200008158:Device{:Zephyr
HXM200008158:78:A5:04:81:AF:F9::0:1}:Zephyr HXM200008158:DataType
    {com.google.heart_rate.bpm[bpm(f)]}}
```

The *Hello Fit* App used the *BleApi* and the *SensorsApi* and used classes from the *data*, *request*, and *result* packages. Data points from the heart rate monitor—at the time of writing—failed to be detected by the App because my attempt to claim the device failed. *OnDeviceFound* appears to be the appropriate place to attempt claiming the device. All the same, I also tried invoking the claim call in the *onScanStopped* callback by retaining the *bleDevice* instance as a member variable in the Activity; but that did not help either.

8.10 Google's Fit App

The *Google Fit App* from Google is available on the Google Play Store at

https://play.google.com/store/apps/details?id=com.google.android.apps.fitness.

Among other features and functionality, the *Google Fit App* lists the other *Google Fit* Apps that connect to the *Fit Store* and denotes them as "third-party, connected" Apps. More details of the *Google Fit* App can be understood by installing the App on your device and studying it thoroughly. The role of the *Google Fit App* does not appear at this time to be central to commencing Fit development. This could change over time so stay tuned.

8.11 Google Settings App

Google Settings—which is an App available on Android 5 devices with Google Play Installed—helps you manage settings for the Google services and applications. Particularly, the Google Settings App provides access to *Google Fit* settings. More information on Google Settings and Managing your *Google Fit* settings is available at

https://support.google.com/accounts/answer/3118621?hl=en&ref_topic=3100928
https://support.google.com/accounts/answer/6098255

References and Further Reading

https://developers.google.com/fit/overview
http://en.wikipedia.org/wiki/Google_Fit
https://developers.google.com/fit/branding
https://developers.google.com/fit/terms
http://developer.android.com/google/play-services/setup.html
https://developers.google.com/fit/android/get-started
https://developer.android.com/tools/projects/index.html

Chapter 9 Google Fit API

9.1 Google Fit API

The *Google Fit API* (also referred to simply as the *Fit* API) resides in the namespace *com.google.android.gms.fitness* and is available as part of *Google Play Services*. Any *Google Fit* App has a dependency on the *Google Play Services* App being installed on the device that it is running on.

9.2 Google fit main package (*com.google.android.gms.fitness*)

The *Google Fit API* documentation is available at https://developers.google.com/fit/android/reference and is a useful reference when reading this chapter.

In all, the main package *fitness* and its sub-packages contain about eight high-level interfaces and 30 high-level classes. The main package *fitness* contains four sub-packages: *data*, *service*, *request*, and *result*. The *fitness* package itself contains the interfaces *SensorsApi*, *BleApi* (Bluetooth Low Energy API), *SensorsApi*, *RecordingApi*, *HistoryApi*, and *SessionsApi*, as well as the classes *Fitness*, *FitnessActivities*, and *FitnessStatusCodes*.

Figure 9-1 shows the classes and interfaces within the main *fitness* package, as well as the sub-packages *data*, *request*, *result*, and *service*.

Wearable Android™: Android Wear & Google Fit App Development, First Edition. Sanjay M. Mishra.
© 2015 John Wiley & Sons, Inc. Published 2015 by John Wiley & Sons, Inc.

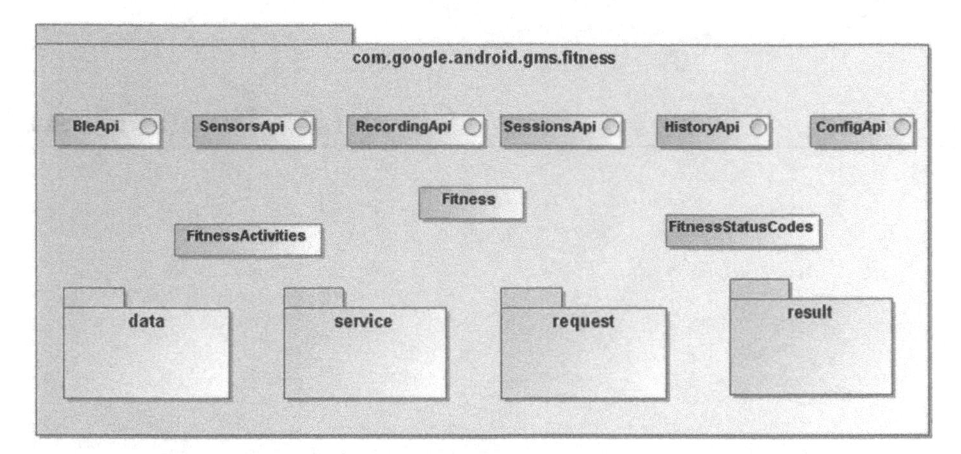

Figure 9-1 Google Fit's fitness package, high-level overview.

9.2.1 *Fitness* class

The *Fitness* class is the entry point into the *Google Fit* API. In order for a *Fit* App to connect to the *Fit* API, a Google user account is required on the device. The *Fit* App needs to specify the scope depending on the data and mode of access desired by the App. The *GoogleApiClient* uses the user account and scope information to get the *OAuth* tokens on behalf of the App, which ensures that only Apps with the necessary permissions can access specific data in the right mode of read or write—after explicit user consent.

An instance of *Fitness* is obtained via using the *GoogleApiClient*'s Builder, adding the Fitness API, specifying the Google account, and adding the scope. The following snippet in the user interface Activity's *onCreate()* method shows how the *Activity* can connect the *GoogleApiClient* with the *Fitness* API added and enabled:

```
googleApiClient = new GoogleApiClient.Builder(this)
                        .addApi (Fitness.API)
                        .useDefaultAccount()
                        .addScope( new Scope (Scopes.FITNESS) )
                        .addOnConnectionCallbacks(this)
                        .addOnConnectionFailedListener(this)
                        .build();
googleApiClient.connect();
```

You will notice that the *Google Fit* App needs to add the "Fitness" API in order to access *Google Fit*. The main package is named *fitness*, and the *Fitness* class is the entry point into the *Google Fit* API. Therefore, the *Fit* API is sometimes also referred to as the *Fitness* API.

The user interface *Activity* that needs to access the *Fit API* will typically need to implement several listeners such as the *GoogleApiClient*'s *ConnectionCallbacks* and *OnConnectionFailedListener* as well as *OnDataPointListener* from the *Fit* API.

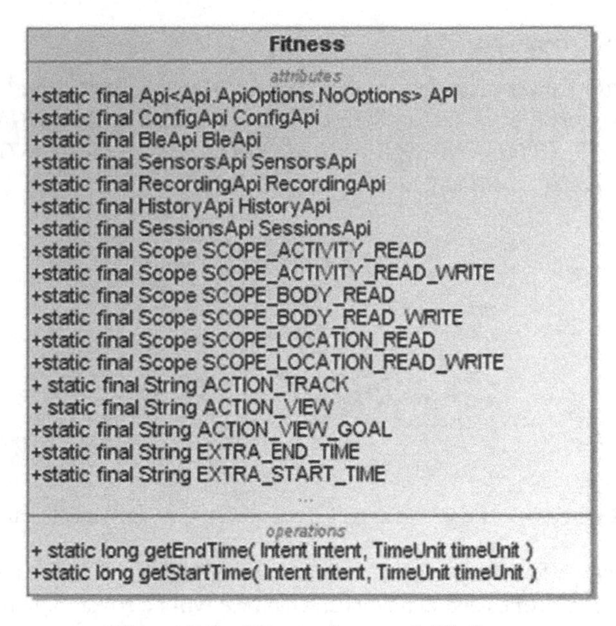

Figure 9-2 Fitness class, partial listing.

Figure 9-2 shows a partial listing of the attributes and methods available in the Fitness class.

At runtime, the *Fit API* will become available only after the *GoogleApiClient* instance has connected successfully. After connecting successfully, the various specific APIs within the Fitness API such as *BleApi*, *SensorsApi*, *SessionsApi*, *RecordingApi*, *HistoryApi*, and *ConfigApi* can all be accessed via the *Fitness* class.

The *Fitness* class defines several ACTION families of *Intent*s as static attributes that support collaboration between different *Fit* Apps. The *ACTION_VIEW* defines an *Intent* to view fitness data, while the *ACTION_TRACK* pertains to tracking a fitness activity; the *ACTION_VIEW_GOAL* defines an *Intent* to view a fitness goal. Associated with these actions/*Intent*s are several extras, that is, *Bundle*s of additional information that can be accessed via the call to *Intent.getExtras()*; a Bundle is a mapping between a key and a value. The *Fitness* class defines extras for the start time (*EXTRA_START_TIME*) and end time (*EXTRA_END_TIME*). Based on the particular action/*Intent*, there are particular relevant extras. The action *ACTION_VIEW_GOAL*, for instance, has the attribute of mimeType, the *EXTRA_START_TIME* and *EXTRA_END_TIME* and *DataSource.EXTRA_DATA_SOURCE*. Because these *Intent*s are intended for collaboration between Apps, it is important to standardize on the names of extras using static final attributes. When you use extras with Intents within the same app too, it is a good programming practice to define the value of the key as a static final constant so that different components within your App can access the extra using the static final constant, rather than hard-coded values, which can be error prone.

The *Fitness* class also defines several *SCOPE*s for physical activity, body, and location, in read and write mode.

9.2.2 FitnessActivities class

The *FitnessActivities* class provides an elaborate set of *public static final String* constants that denote real-world physical activities such as *BASEBALL*, *BIKING*, *RUNNING*, *WALKING*, *TENNIS*, *SWIMMING*, *SQUASH*, *MARTIAL_ARTS*, *DANCING*, *GARDENING*, *SLEEP*, *MEDITATION*, and many more.

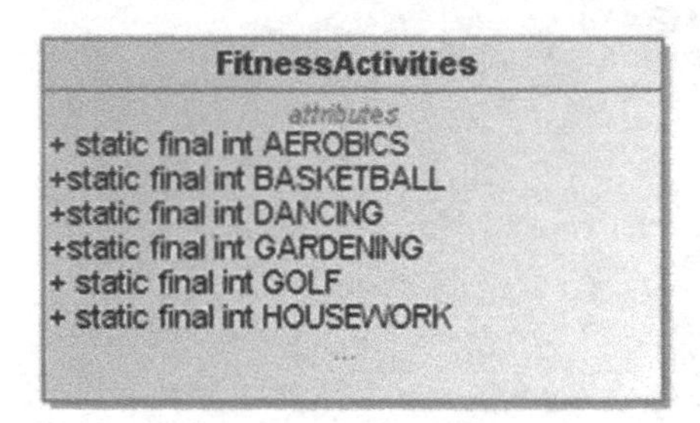

Figure 9-3 FitnessActivities class, with a few constants listed.

Figure 9-3 shows a few of the constants representing physical activities and sports. These constants are used in *Sessions* and *DataTypes*, which we will be covering in the next sections.

9.2.3 FitnessStatusCodes class

The *FitnessStatusCodes* class defines about 18 constants that represent the result of requests made to the *Google Fit* API. The *FitnessStatusCodes* help in pinpointing the exception, error, or conflict when making a request to the *Google Fit* API. The *request* sub-package, which is covered in the next few sections contains various types of requests pertaining to Bluetooth LE (Ble), sensors, sessions, and so on. The *result* sub-package contains various types of corresponding results, which encapsulate the result data as well as the status. The App making the request needs to get the status in order to confirm that the request was successful or address the issue before making the attempt again.

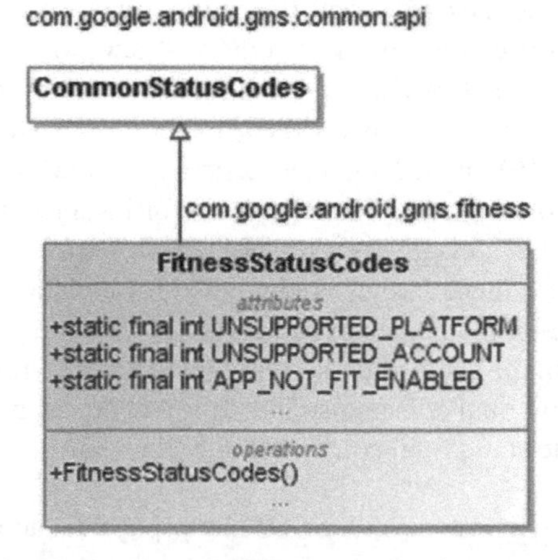

Figure 9-4 FitnessStatusCodes class, with a few constants listed.

Figure 9-4 shows a few of the constants available in the *FitnessStatusCodes* class.

9.2.4 BleApi interface

The *BleApi* is accessible via the static attribute available in the *Fitness* class. Prior to accessing *BleApi*, the *GoogleApiClient* must have connected successfully.

The *BleApi* provides functionality for scanning, claiming, and using Bluetooth Low Energy (Ble) Fitness devices. Many Bluetooth LE devices accept connections readily without the need for pairing. The concept of claiming a fitness device ensures that *Google Fit* apps connect only with fitness devices that the user owns and per the user's intentions. The user needs to explicitly claim a device before it can be used by *Google Fit* in the sequential flow of scan/detect, claim, and use. Once a device is claimed, its data sources become available via the *SensorsApi* and the *RecordingApi*. Also, the App should use *Google Fit* to connect to the device rather than connect directly with the device (by using the base Android API's *android.bluetooth.** sub-packages and classes).

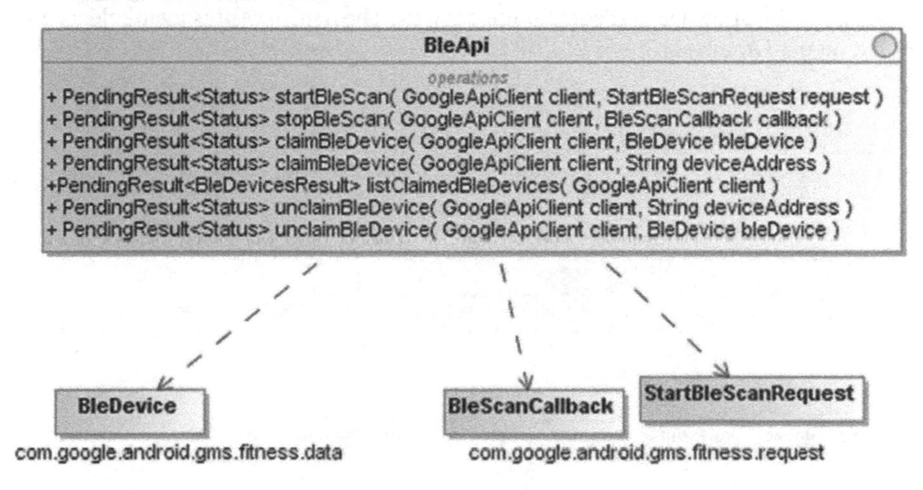

Figure 9-5 BleApi.

Figure 9-5 shows the *BleApi* interface with its methods related to scanning, claiming, and listing Bluetooth LE devices.

The scan process is closely associated with *StartBleScanRequest* and the *BleScanCallback* both of which reside in the *request* package and have been covered in more detail in the next sections.

The following code snippet shows the creation of a scan request by using the *StartBleScanRequest* and the *BleScanCallback* classes. There must be at least one data type associated with the *StartBleScanRequest*:

```
PendingResult <Status> r = Fitness.BleApi.startBleScan (
        googleApiClient, new StartBleScanRequest.Builder()
                        .setDataTypes (DataType.TYPE_HEART_RATE_BPM)
                        .setBleScanCallback( bleScanCallback )

                        .build()) ;
```

The scan is asynchronous, and the *BleScanCallback* supports finding devices via the scan. Devices that are found can be claimed; however, any active Bluetooth scan operations should be stopped prior to claiming a device. The *PendingResult* has callbacks that can help the App determine the success or failure of the call.

The *BleApi* requires that Bluetooth is enabled in order for its methods to work. The *FitnessStatusCode.DISABLED_BLUETOOTH* represents the condition that Bluetooth is not enabled. Apps can address this by using the *startResolutionForResult()* method on the *Status(com.google.gms.common.api)*. This starts the appropriate *Intent* that requires user interaction to resolve the condition. The *PendingResult*, which is used in all the *BleApi* method calls, is used in calls to the *Google Play Services* API.

9.2.5 SensorsApi

The *SensorsApi* provides access to live, real-time streams of sensor data from hardware sensors on the local device and companion devices. The *SensorsApi* is available as a static attribute on the *Fitness* class.

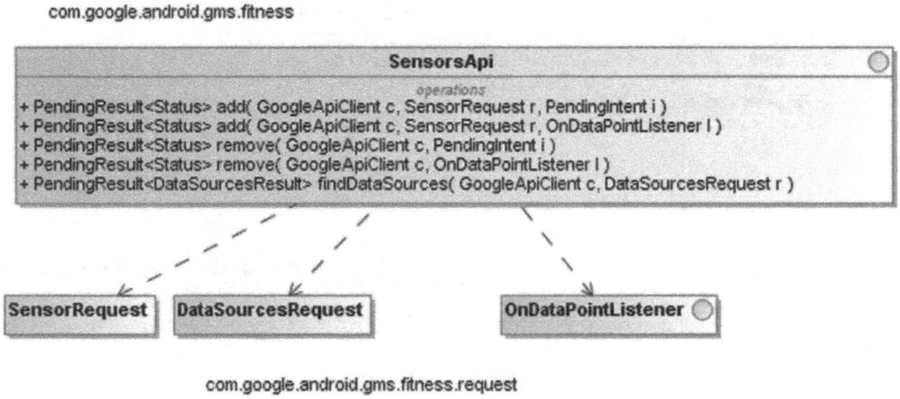

com.google.android.gms.fitness

SensorsApi

operations

+ PendingResult<Status> add(GoogleApiClient c, SensorRequest r, PendingIntent i)
+ PendingResult<Status> add(GoogleApiClient c, SensorRequest r, OnDataPointListener l)
+ PendingResult<Status> remove(GoogleApiClient c, PendingIntent i)
+ PendingResult<Status> remove(GoogleApiClient c, OnDataPointListener l)
+ PendingResult<DataSourcesResult> findDataSources(GoogleApiClient c, DataSourcesRequest r)

SensorRequest

DataSourcesRequest

OnDataPointListener

com.google.android.gms.fitness.request

Figure 9-6 SensorsApi.

Figure 9-6 shows the methods available in the *SensorsApi* interface. The *SensorsApi* provides methods to add and remove listeners for sensor data. The add method whose parameters include a *PendingIntent* parameter is useful for slower sampling of sensor data. This method adds a *PendingIntent* listener to a sensor data source. Once the call to add succeeds, the *PendingIntent*'s callback will provide access to new sensor data, every time new data arrives. The application can extract the *DataPoint* (which encapsulates a sensor value and a timestamp) from the intent. *DataPoint* resides in the *data* package, which we will be covering shortly. There is another flavor of the add method whose parameters include an *OnDataPointListener*. This method is more appropriate for faster sampling rates using a live listener in the foreground. After the add succeeds, new *DataPoints* in the data stream are delivered to the specified listener.

The following snippet shows an example of adding a sensor request:

```
PendingResult <Status> r = Fitness.SensorsApi.add ( googleApiClient,
                          new SensorRequest.Builder()
                          .setDataType (DataType.TYPE_HEART_RATE_BPM )
                          .setSamplingDelay ( 2, TimeUnit.MINUTES )

                          .build(), myOnDataPointListener );
```

The listeners need to be removed when the UI *Activity* is paused, stopped, or destroyed; and added again when the UI *Activity* resumes or restarts. The *remove* methods are used for removing the listener from the sensor data source. The *findDataSources()* method is used to find all available data sources for the specified data types. It is not necessary to call this method if your application is interested in getting the best available data of the specified data type, irrespective of the source, which may often be the case.

9.2.6 RecordingApi

The *RecordingApi* supports the collection, that is, recording of sensor data into the *Google Fit Store* in the background, in a low power consuming, always-on mode. The *RecordingApi* provides methods to subscribe to a given data type. Subscriptions persist device restarts and work in the background, irrespective of whether the subscribing App is running or not. Subscribing to a data type requires the user's permission and consent, and the user can subsequently revoke such permission via the *Google Play Services*' settings.

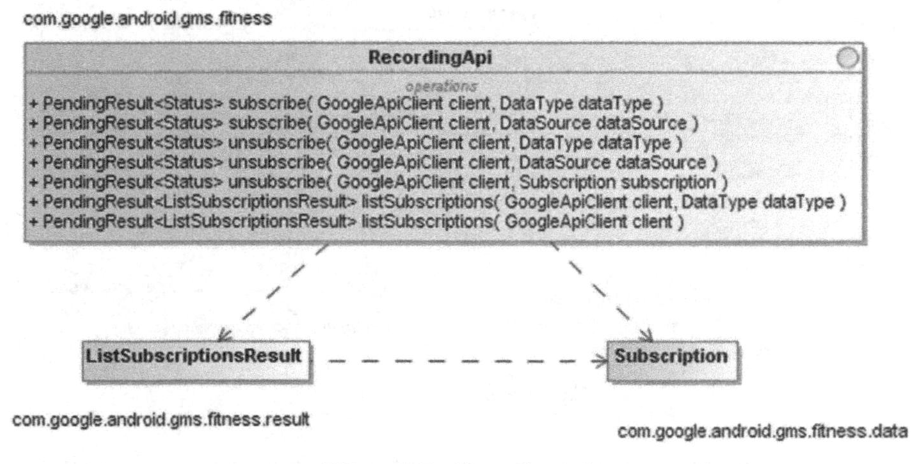

Figure 9-7 RecordingApi.

Figure 9-7 shows the methods in the *RecordingApi*, which include *subscribe*, *unsubscribe*, and *listSubscriptions.*

The following code snippet shows the use of the SensorsApi to subscribe to a data type:

```
PendingResult <Status> r  = Fitness.RecordingApi.subscribe ( googleApiClient ,
                              DataType.TYPE_HEART_RATE_BPM ) ;
```

The *RecordingApi*, like the *SensorsApi*, becomes available only after a device has been successfully claimed via the *BleApi*. Unlike the *SensorsApi*, the *RecordingApi* does not deliver any live sensor data to the application. An App can have both a *SensorsApi*-based listener running and a *RecordingApi*-based subscription at the same time.

9.2.7 SessionsApi

The *SessionsApi* is useful for creating and managing sessions of user's physical activity. A session represents a time interval during which a user engages in physical activity. A *Session* has a user-readable name such as "Morning Run," a start time and associated data that is stored in the *Fit Store*. All the data within the time range is implicitly associated with the session. Session data is stored in a shareable manner and can be queried.

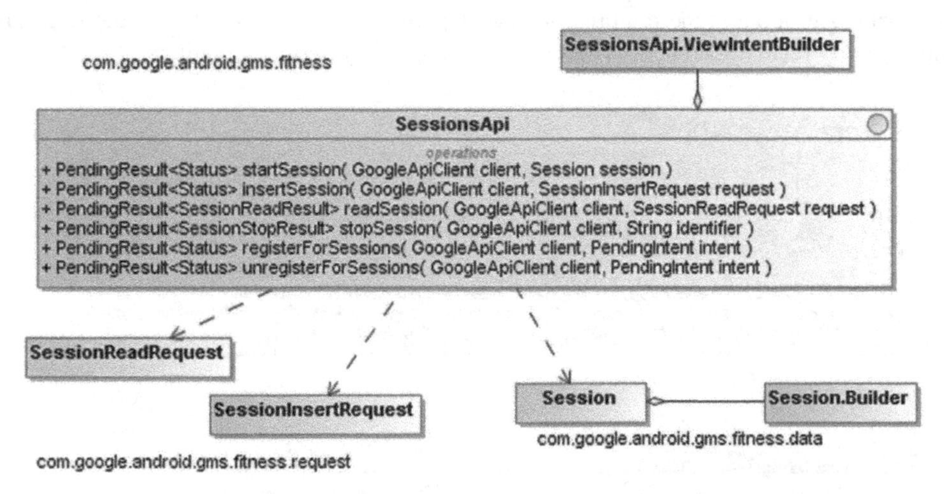

Figure 9-8 SessionsApi.

Figure 9-8 shows the methods in the *SessionsApi* interface. Once a session is started using the *startSession* method, session data can be stored in the *Google Fit Store* using the *insertSession* method using a *SessionInsertRequest*; also, any data that is inserted using the *HistoryApi* during the duration of the session will be associated with the session. Thus, *insertSession* and the *HistoryApi* methods represent two ways by which session data can be stored into the *Google Fit Store*.

Once session data has been stored in the *Google Fit Store*, it can be retrieved by using the *readSession* method. The *stopSession* will terminate the session. The session also terminates upon the end of its duration when its end time has been reached. The *register-ForSessions* method allows the application to be notified of session's start and end events, via the *PendingResult*. The inner class *SessionsApi.ViewIntentBuilder* is useful for displaying detailed session data stored in the *Google Fit Store*.

The setting of the start time for the session is mandatory, while the end time is optional. The *stopSession* method available in the *SessionsApi* allows an application to stop a session. The following code snippet shows the use of the *SessionsApi* to start a new session of 20 minute duration:

```
long startTime = System.currentTimeMillis() ;
long endTime = startTime + 20*60*1000 ;
Session joggingSession = new Session.Builder()
                        .setName("Morning Jog")
                        . setStartTimeMillis ( startTime,
                                TimeUnit.MILLISECONDS )
                        .setEndTime ( endTime)
                        .build() ;
PendingResult <Status> r = Fitness.SessionsApi.startSession ( googleApiClient,
                                joggingSession ) ;
```

After your App has created a session, you can insert the session and its associated data into the *Google Fit Store*. Your App can also read session data subsequently. There are Session Start and End intents that your App can register a broadcast listener for, in order to handle session starts and ends.

9.2.8 HistoryApi

The *HistoryApi* supports inserting, reading, and deleting data in the *Google Fit Store*. The *HistoryApi* also supports insertion of data that was collected outside of *Google Fit*, which can be useful if some readings were entered manually by the user or imported from a device that is not supported by *Google Fit*. The *HistoryApi* enables your App to perform bulk operations on the fitness store: inserting, deleting, and reading fitness data.

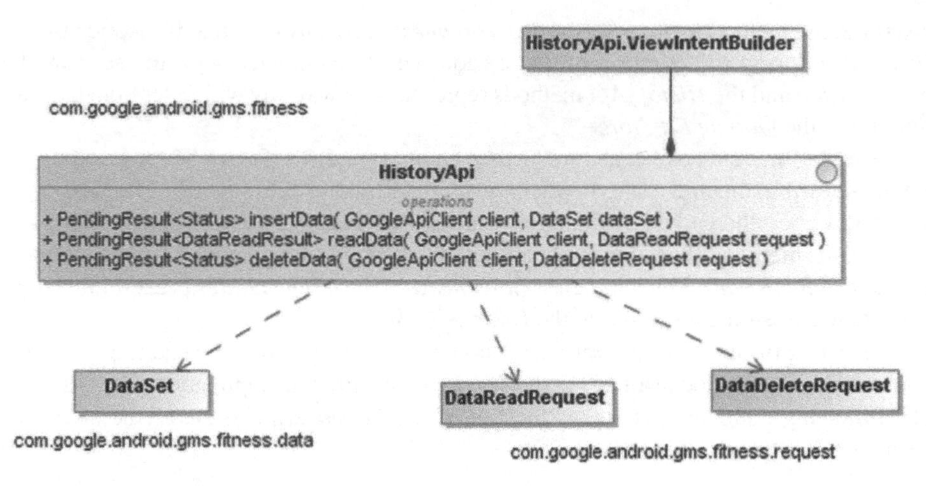

Figure 9-9 HistoryApi.

Figure 9-9 shows the methods available on the *HistoryApi* interface. The *insertData* method is useful for inserting data that was collected outside of *Google Fit*, including data entered manually by the user and/or bulk data. The *readData* method is useful for reading historical data.

9.2.9 ConfigApi

The *ConfigApi* is useful for accessing settings in *Google Fit* as well as creating and accessing custom data types.

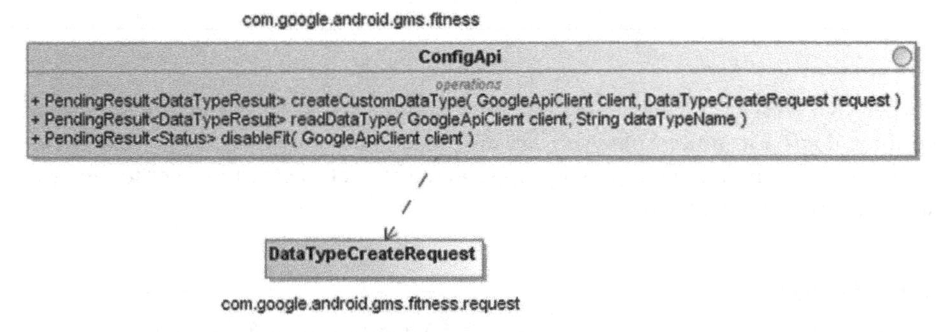

Figure 9-10 ConfigApi.

Figure 9-10 shows the methods available in the *ConfigApi*. The *readDataType* method is useful for retrieval of shareable data types added by other App or a custom data type added by your App. The *createCustomDataType* supports making a request to create a new data type and adding it to the *Google Fit* platform. The *disableFit* method supports disabling the App from *Google Fit*. Apps should provide users with the *Disconnect from Google Fit* option in the App's settings. Disconnecting from *Google Fit* revokes the OAuth permissions and removes all the sensor registrations and recording subscriptions.

9.3 data sub-package

The data sub-package has the fully qualified package name of *com.google.android.gms.fitness.data* and has about 11 classes (Figure 9-11A).

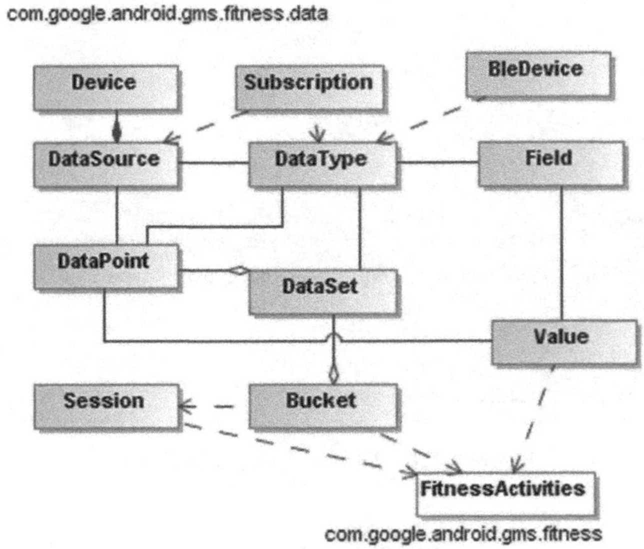

Figure 9-11A data package.

9.3.1 Device

The *Device* class represents an integrated device such as a handheld device or an *Android Wear* device that can hold sensors. The *Device* class encapsulates the manufacturer and model information, which can help in identifying the source of sensor data and distinguishing between two similar sensors such as heart rate monitors on two different devices. The *Device* class is also useful in distinguishing between the data patterns from similar sensors on different types of devices such as accelerometer data from a smart watch versus a handheld device.

Figure 9-11B shows some of the main attributes and methods available in the *Device* class. There are several constants for the *Device* type such as *TYPE_WATCH*, *TYPE_CHEST_STRAP*, *TYPE_SCALE*, *TYPE_PHONE*, *TYPE_TABLET*, and *TYPE_UNKNOWN*.

9.3.2 BleDevice

The *BleDevice* class represents a Bluetooth LE device that advertises information about its onboard sensors (such as heart rate monitor, step counter, and so on).

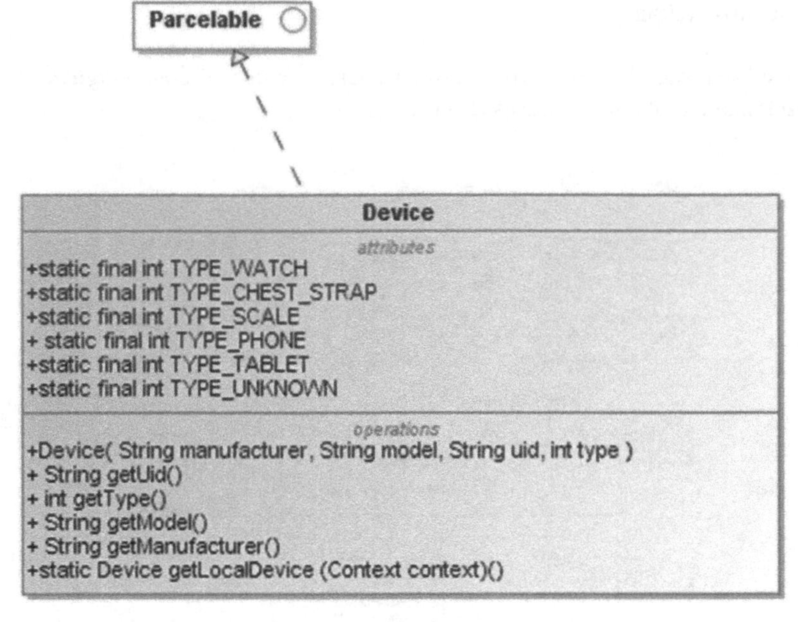

Figure 9-11B Device class, partial listing.

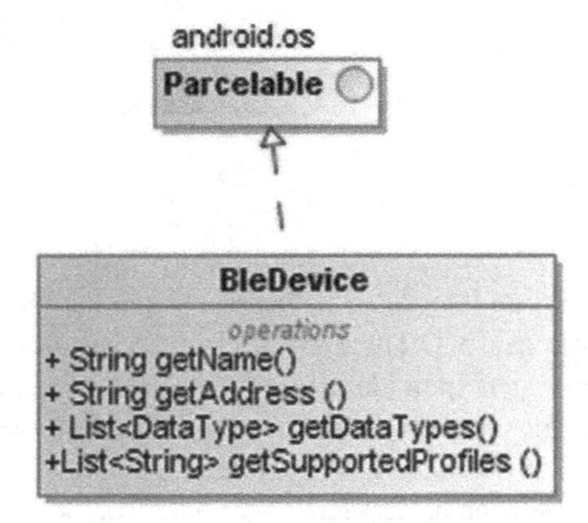

Figure 9-11C BleDevice class, partial listing.

Figure 9-11C shows the *BleDevice* class that has methods for getting the device name, the *Fit* data types supported by the device, and the Bluetooth *Generic Attribute Profile* (GATT). More information can be found at

https://developer.bluetooth.org/gatt/profiles/Pages/ProfilesHome.aspx and
https://developer.bluetooth.org/gatt/Pages/GATT-Specification-Documents.aspx.

9.3.3 DataSource

A *DataSource* represents a unique source of sensor data and can expose the raw data that's coming from a particular hardware sensor on the host device or a peripheral device. A *DataSource* can also expose data derived from merging or transforming data from other data sources.

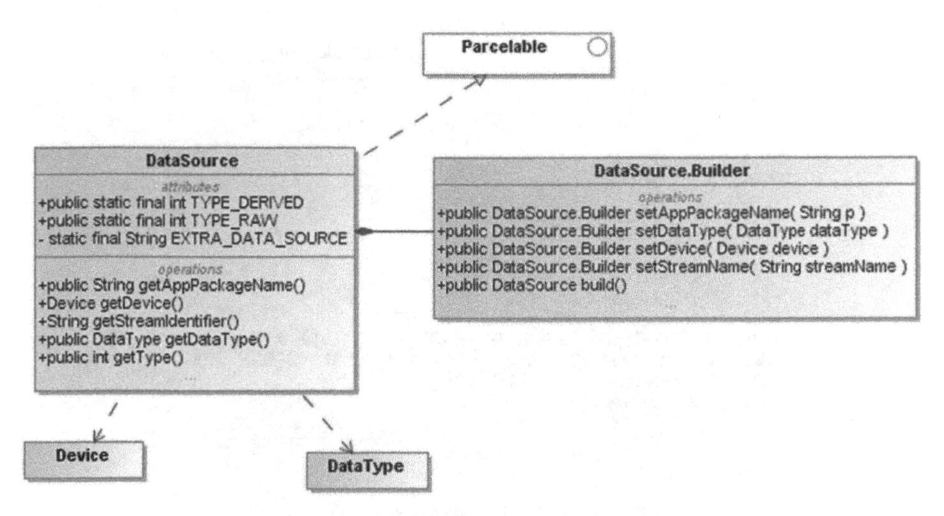

Figure 9-11D DataSource class, partial listing.

Figure 9-11D shows the methods and attributes available in the *DataSource* class. A *DataSource* instance contains information that uniquely identifies it such as the hardware device and the App that collected or transformed the data. There can be multiple *DataSource*s for the same *DataType*, which mirrors the reality that there can be multiple sensors for the same data types and fields such as heart rate, step counters, and so on.

Apps can access the data stream from a *DataSource* in near real time by registering an *OnDataPointListener* or via making queries at periodic intervals.

9.3.4 DataType

The *DataType* class defines the representation of the data, which is independent of how the data was acquired, the sensor that was used, and so on. The *DataType* is the schema for a stream of data that can be collected and inserted into and queried from the *Fit Store*. A *DataType* contains one of more fields. Each field has a name and a format. The format indicates whether it is an int or a float denoted by *FORMAT_INT_32* and *FORMAT_FLOAT*, respectively.

Each *DataType* has a unique namespaced name, for example, *com.google.heart_rate.bpm* is the namespaced name for the Java attribute *TYPE_HEART_RATE_BPM* defined in the *DataType* class (Figure 9-11E).

9.3.5 DataPoint

A *DataPoint* holds at least one field and a value representing a single data point in a data type's stream from a particular data source (Figure 9-11F). Every *DataPoint* that is stored to or retrieved from *Google Fit* has an associated *DataSource*. Each *DataSource* in turn contains information of the sensor device and/or App that has collected or transformed the data.

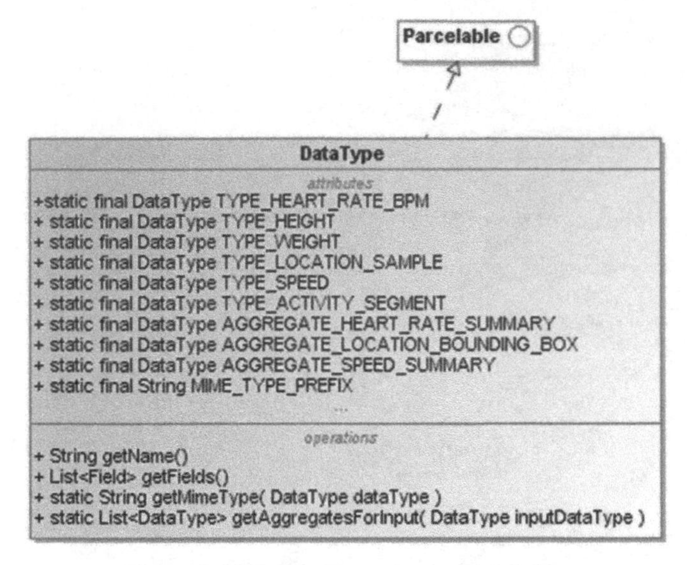

Figure 9-11E DataType class, partial listing.

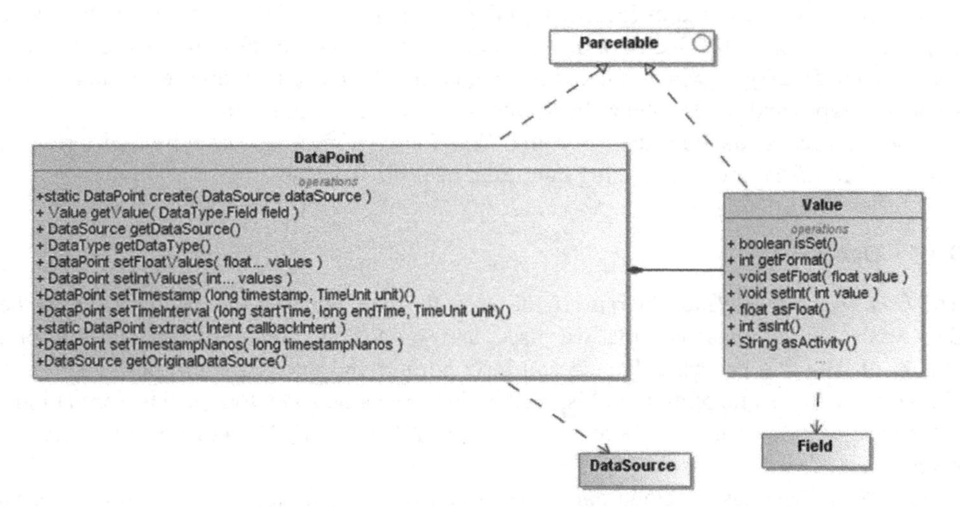

Figure 9-11F DataPoint class, partial listing.

Each *DataPoint* holds at least one field and the value of data, as well as a timestamp. Optionally, a Data point may also contain a start time. The exact semantics and combination of field data depends on the particular data type. Some data points represent an instantaneous reading, such as location, speed, and heart rate beats per minute. Other data points represent aggregate data, such as heart rate summary.

DataPoint instances are created by making a call to the static method DataPoint.create(), which requires a data source as the method call argument. Recall that each data source has an associated data type and each data type has associated fields.

When a *DataPoint* is instantiated, it contains all the appropriate *l-values* for the fields associated with the data type, but the r-values (values of data or content) have not been set. In other words, initially, all the "l-values" are present and the appropriate r-values and timestamp will need to be set. "*l-values*" represent the variable name, while "*r-values*" represent the content or data that's set into a given variable (*l-value*). The concept of *l-values* and *r-values* is used in computer science as well as in programming languages.

As an illustration of the concept of l-values and r-values, the l-values in my shell environment are *ANDROID_HOME*, *ANT_HOME*, *JAVA_HOME*, and so on, while the r-values are the data that these variables have been set to such as */opt/androidsdk/sdk*. l-values can exist with empty r-values (eg. NOTHING_HOME):

```
$ env | grep HOME
    ANDROID_HOME=/opt/androidsdk/sdk
    GRADLE_HOME=/opt/tools/gradle
    ANT_HOME=/opt/tools/ant
    MAVEN_HOME=/opt/tools/maven
    JAVA_HOME=/opt/jdk1.7
    HOME=/home/sanjay

    NOTHING_HOME=
```

The snippet below shows the setting of the *r-value* of a field in a *DataPoint* instance:

```
dataPoint.getValue(0).setInt ( anIntValue)
```

The *getDataSource* method returns the *DataSource* for the *DataPoint*. The call to *getOriginalDataSource* returns the original data source, which can—in the case of transformed or merged data—be different from the data source returned by the getDataSource method.

9.3.6 Field

The *Field* class represents one dimension of a data type and has a name and a format. The data type *DataType.TYPE_LOCATION_SAMPLE*, for example, has four fields: latitude, longitude, altitude, and accuracy.

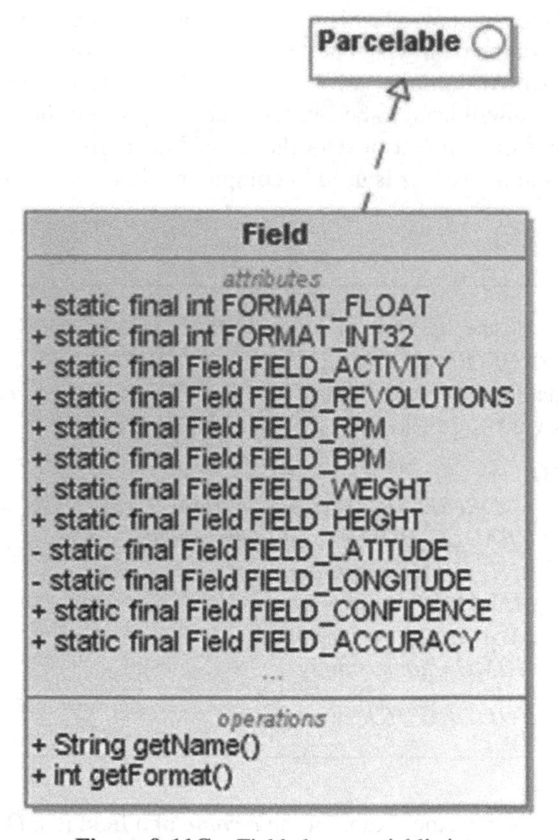

Figure 9-11G Field class, partial listing.

Figure 9-11G shows the Field class with some of the static constants defined therein. The Field names are not namespaced, they are unique within the data type. Some fields such as *FIELD_CONFIDENCE* and *FIELD_ACCURACY* are secondary to their primary accompanying Field.

9.3.7 Value

The *Value* class is a holder for a single field in a *DataPoint*. Depending on the *DataType* of the *DataPoint*, a *Value* instance is created for each Field, but it has not been set with the data (r-value). The *isSet* method indicates whether a value has been set on the *Value* instance. Figure 9-11F shows the *Value* class along with the *DataPoint*.

9.3.8 Subscription

The *Subscription* class encapsulates the data source and data type of a subscription (Figure 9-11H). The concept of subscription is associated with background collection of sensor data, which we covered in the section on the *RecordingApi*.

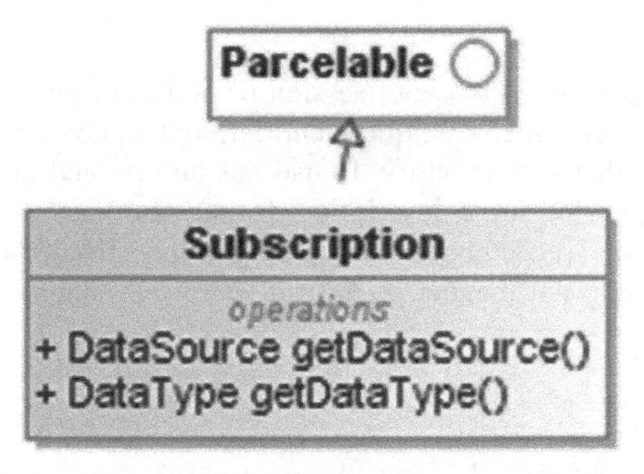

Figure 9-11H Subscription class.

The *Subscription* class is not instantiated directly, rather the calls to the *RecordingApi*'s *subscribe* and *listSubscriptions* method and associated callbacks provide access to the *Subscription* instances that are contained within the *ListSubscriptionsResult* class.

9.3.9 DataSet

The *DataSet* represents a fixed set of *DataPoint*s in a *DataType*'s stream from a particular *DataSource*. Typically, a *DataSet* contains *DataPoint*s at fixed interval boundaries. A *DataSet* can be used both for batch data insertions and for reads (Figure 9-11I).

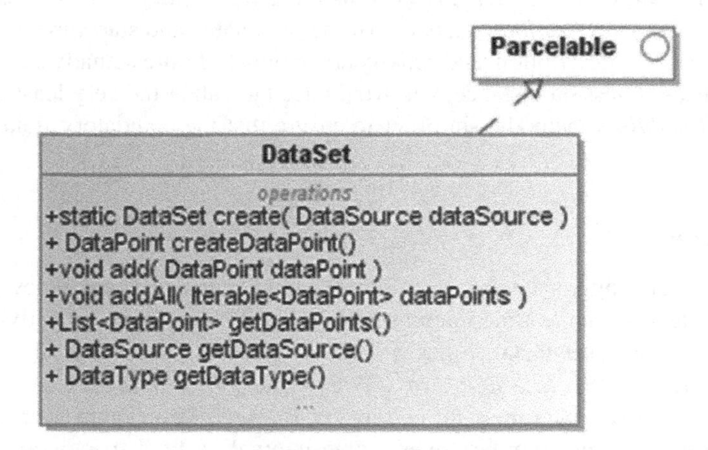

Figure 9-11I DataSet.

The *DataSet* class has a static *create* method that needs a *DataSource* instance as the parameter and returns an instance of *DataSet*. Once you have an instance of *DataSet*, calling the *createDataPoint* method yields an instance of *DataPoint*. After setting values on the *DataPoint* instance, the *DataSet*'s *add* method facilitates adding a data point to the set. The *addAll* method facilitates adding multiple data point instances in one call.

9.3.10 Session

The *Session* class represents a workout session or period of physical activity—it has a user-visible name, a start time, a unique identifier, and some associated data. It has an associated start time that is mandatory. It also has an optional description that can be used by the user to provide some description or notes associated with the session. The *Session* supports storing, retrieving, and analyzing user-visible groups of data organized and aggregated in a relevant manner. A *Session* can be instantiated via the static inner Builder class.

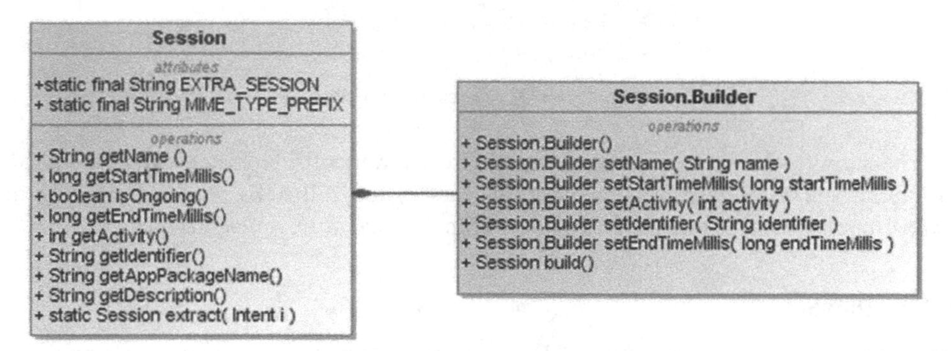

Figure 9-11J Session class.

Figure 9-11J shows the *Session* class and its inner static *Builder* class. *Session.Builder* supports the creation of *Session* instances. The session name and start time are mandatory fields. The identifier, description, and activity are optional. Before actually calling the *build* method to obtain a Session instance, you would need to call at the very least the *setName* and *setStartTimeMillis* methods—in order to ensure that the mandatory fields have been populated.

9.3.11 Bucket

A *Bucket* represents aggregate data over an interval of time using one of several possible bucketing strategies such as time interval, activity type, a session or an activity segment. Accordingly, any Bucket instance has a bucket type that can be one of *TYPE_TIME*, *TYPE_ACTIVITY*, *TYPE_SESSION*, *or TYPE_ACTIVITY_SEGMENT.* A *Bucket* must have a start time and an end time—for all types of Buckets. A Bucket may be setup to contain speed and heart rate summary over a time interval. A Bucket may coincide with a session, via using the bucketing strategy of *TYPE_SESSION*, but that does not always have to be the case.

Figure 9-11K shows the *Bucket* class that has no constructor or builder class. The *Bucket* is not instantiated by application code directly; rather, the *DataReadRequest* in the *request* sub-package has a *bucketBy** family of methods in its *Builder* class, which allow your application to specify the particular criteria for computing the bucket. In turn, the *DataReadResult* from the *result* package has a *getBucket*, which facilitates access of bucket instances from the application code.

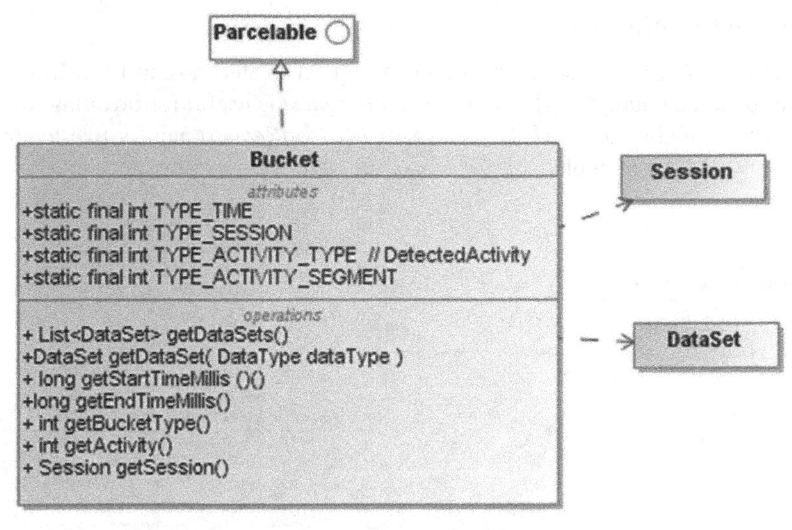

Figure 9-11K Bucket class.

9.4 request sub-package

The *request* sub-package has the fully qualified package name of *com.google.android.gms.fitness.request* and contains about nine classes and one interface.

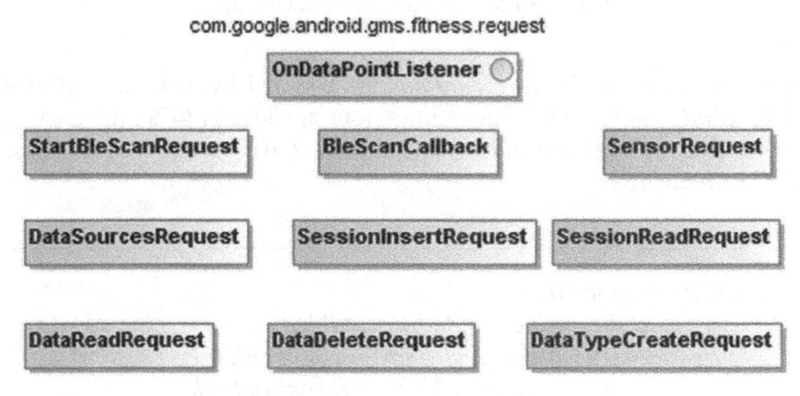

Figure 9-12A request sub-package.

Figure 9-12A shows the various requests, callback, and listener available in the *request* sub-package.

The *Google Fit* API calls pertaining to Bluetooth LE scans, sensors, data sources, sessions, and so on follow an asynchronous, nonblocking mode of calls. The *request* sub-package has a complementary *result* sub-package, and individual request classes from the *request* sub-package typically have a corresponding result class in the *result* sub-package. Both the *request* and *result* sub-packages depend on the *data* sub-package.

9.4.1 StartBleScanRequest

The *StartBleScanRequest* class encapsulates a request to start a scan for a Bluetooth LE device based on the data type. The *StartBleScanRequest* is useful for invoking the *startBleScan* method available in the *BleApi*. The *StartBleScanRequest* and for that matter all the request family of classes implement the *android.os.Parcelable* interface.

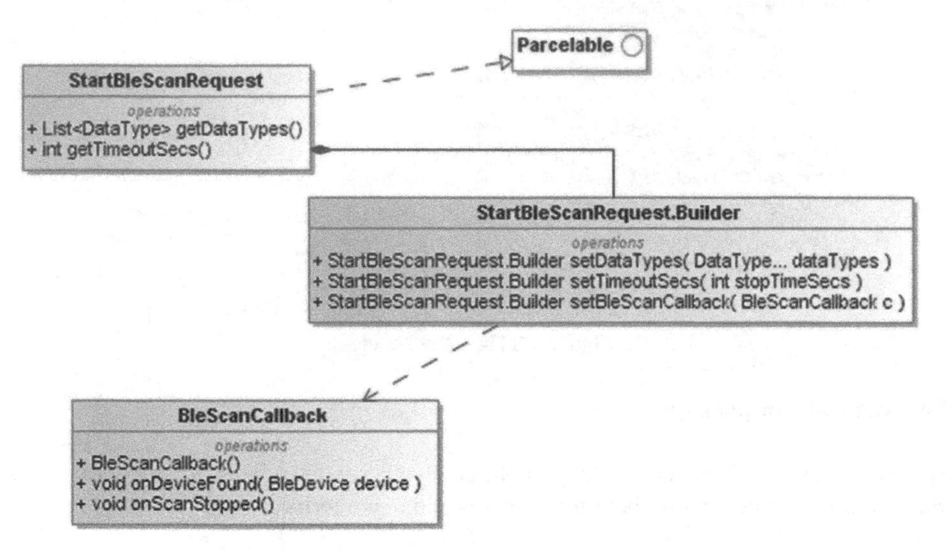

Figure 9-12B StartBleScanRequest class.

Figure 9-12B shows the *StartBleScanRequest* class and its static inner *Builder* class. The *StartBleScanRequest.Builder* class allows your application to set the data types, the scan timeout, and the *BleScanCallback* instance in order to build an instance of the *StartBleScanRequest*:

```
StartBleScanRequest myScanRequest =
                    new StartBleScanRequest.Builder()
                    .setDataTypes( DataType.TYPE_HEART_RATE_BPM )
                    .setBleScanCallback ( myBleScanCallback )
                    .build() ;
```

After you have created an instance of the StartBleScanRequest, you will typically invoke the *startBleScan* method in the *BleApi*, as shown in the snippet below:

```
PendingResult <Status> r = Fitness.BleApi.startBleScan (
                                googleApiClient, myScanRequest ) ;
```

9.4.2 BleScanCallback

The *BleScanCallback* is an abstract class that is associated with the *BleApi* and the *StartBleScanRequest*. Figure 9-12B shows the BleScanCallback and its abstract methods. Your application will need to extend this abstract class and implement the *onDeviceFound* and the *onScanStopped* methods.

9.4.3 SensorRequest

The *SensorRequest* class is used to request real-time data of a particular data type from a particular data source. It also lets you specify the accuracy mode, sampling rate, fastest reporting rate, delivery latency, and so on. Greater accuracy and more frequent sampling generally come at the cost of higher power consuming and faster draining of the battery. The *SensorRequest* class provides constants for the accuracy mode.

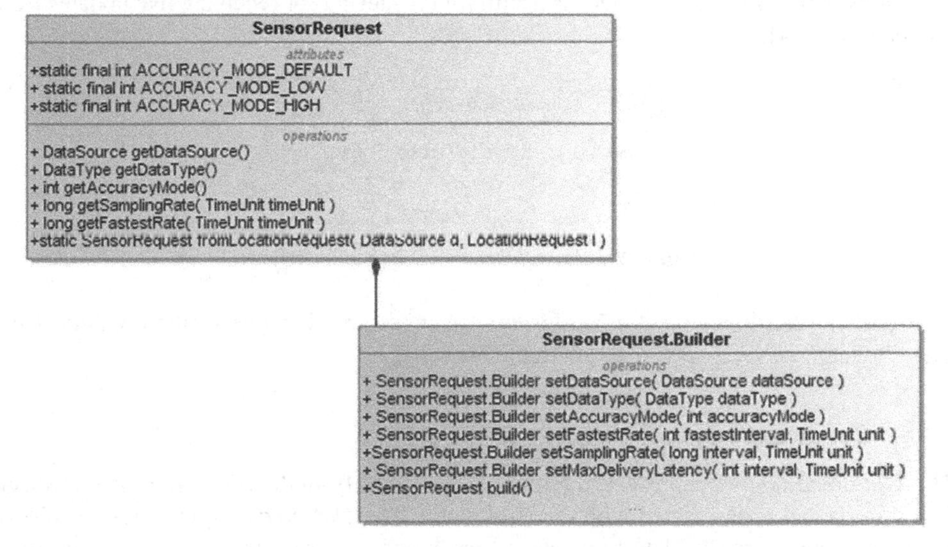

Figure 9-12C SensorRequest class.

Figure 9-12C shows the attributes and methods of the *SensorRequest* class and its static inner *SensorRequest.Builder* class, which provides methods to build a *SensorRequest* instance after specifying the data source, data type, accuracy mode, sampling rate, maximum delivery latency, and so on. The delivery latency is the time delay between the time of origination of data at the sensor source and the receipt of the data update within component made the request for the data.

9.4.4 DataSourcesRequest

The *DataSourcesRequest* class represents a request to find *Google Fit* data sources that match specified criteria. Recall that a data source is a source of sensor data that can expose raw data from a hardware sensor or expose derived, transformed, or merged data.

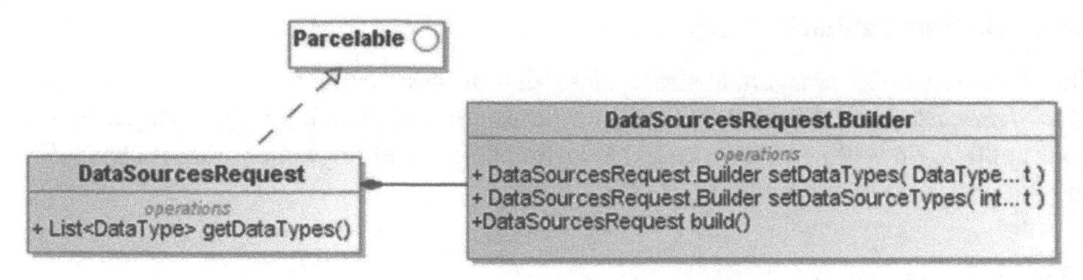

Figure 9-12D DataSourcesRequest class.

Figure 9-12D shows the *DataSourcesRequest* and its inner static *Builder* class. The *DataSourcesRequest* is closely associated with the *SensorsApi*'s *findDataSources* method.

9.4.5 OnDataPointListener

The *OnDataPointListener* interface is useful for registering for receiving live updates from a *DataSource*, which are delivered as *DataPoints*.

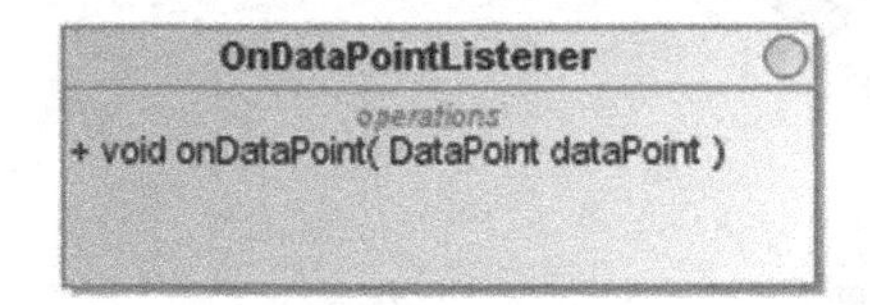

Figure 9-12E OnDataPointListener interface.

Figure 9-12E shows the *OnDataListener* interface that has one method *onDataPoint*. The *OnDataListener* is associated with the *SensorApi*.

9.4.6 DataReadRequest

The *DataReadRequest* represents a request to read data from the *Fit Store* based on specified criteria and is associated with the *HistoryApi*. The *DataReadRequest* must specify at least one data source or data type and a time range. In order to request read access to aggregate data, the request should additionally specify the bucketing strategy. Recall from the section on the *Bucket* that the *Bucket* class defines *Bucket* types of session, activity type, activity segment, and time and contains aggregate data of one or more data types.

Figure 9-12F shows the *DataRequestRequest* class and its static inner *Builder* class, consistent with the pattern we have seen for the various other request family classes.

The snippet below shows an example of creating a simple *DataReadRequest* by setting the time range and data type using the *Builder*:

```
DataReadRequest myReadRequest =
        new DataReadRequest.Builder()
        .setTimeRange ( startMillis, endMillis, TimeUnit.MILLISECONDS )
        .read ( DataType.TYPE_HEART_RATE_BPM )
        .build() ;
```

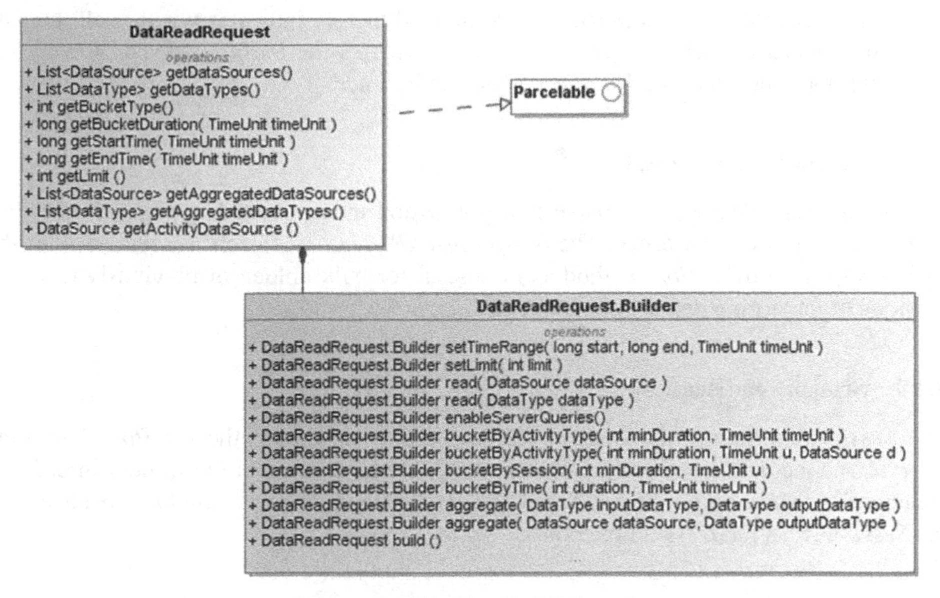

Figure 9-12F DataReadRequest.

The *HistoryApi*'s *readData* method takes in a *DataReadRequest* instance as a parameter.

9.4.7 DataDeleteRequest

The *DataDeleteRequest* is used to specify the criteria for deleting history data. The *DataDeleteRequest* must specify the time interval, and it may be the data type or data source.

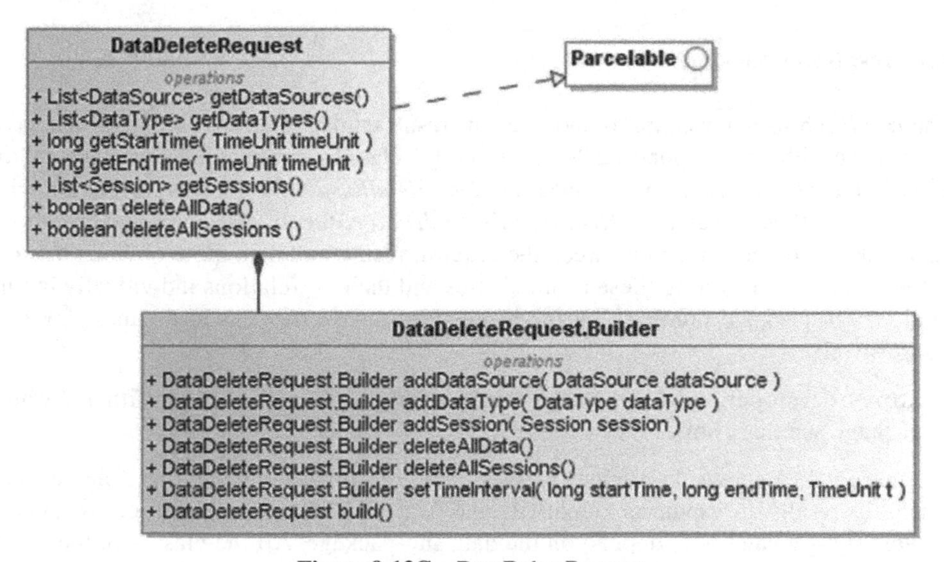

Figure 9-12G DataDeleteRequest.

Figure 9-12G shows the *DataDeleteRequest* and its static inner *Builder* class.

The *DataDeleteRequest's deleteAllData* method merely indicates whether all the data types are marked for deletion. This correlates to whether the *Builder*'s *deleteAllData* was invoked prior to building the *DataDeleteRequest* instance.

9.4.8 SessionInsertRequest

The *SessionInsertRequest* is used to inserting a session and associated aggregate *DataPoint*s or *DataSet*s into the *Fit Store*. The *SessionInsertRequest* is closely associated with the *SessionsApi*'s *insertSession* method and is useful for bulk upload of previously recorded sessions or for storing data from outside of *Google Fit*.

9.4.9 SessionReadRequest

The *SessionReadRequest* is used for reading session data from the *Fit Store*. The time interval and the data types are parameters that can be specified in the Builder in order to create a *SessionReadRequest* instance. The *SessionReadRequest* is closely associated with the *SessionsApi*'s read*Session* method.

9.4.10 DataTypeCreateRequest

The *DataTypeCreateRequest* is used for creating an application specific, custom data type in the *Fit Store*. Such a data type should not duplicate an existing standard/public data type. The data of this custom data type will be private to the App that created it. The data type's name should reside in the namespace of its application's package name. The *DataTypeCreateRequest* is associated with the *ConfigApi*'s createCustomDataType method.

9.5 result sub-package

The *result* sub-package contains about seven "result family" classes, each of which has a correlation with a corresponding "request family" class from the *request* package and/or a particular *Fit* API call. For instance, the *DataReadResult* class has a correlation with the *DataReadRequest* and the *HistoryApi*'s *readData* API call. The naming conventions also make such correlations between the request, result, and data quite obvious in most cases. We will be covering these result classes and their correlations individually in this section. The package summary for the result sub-package happens to list these correlations as well:

https://developer.android.com/reference/com/google/android/gms/fitness/result/package-summary.html

Figure 9-13A shows the various classes in the *result* package, and these classes represent the results or responses to requests and API calls. The *result* sub-package and the *request* sub-package both depend on the data sub-package. All the classes in the *result* package implement the *Result* interface from *com.google.android.gms.common.api*

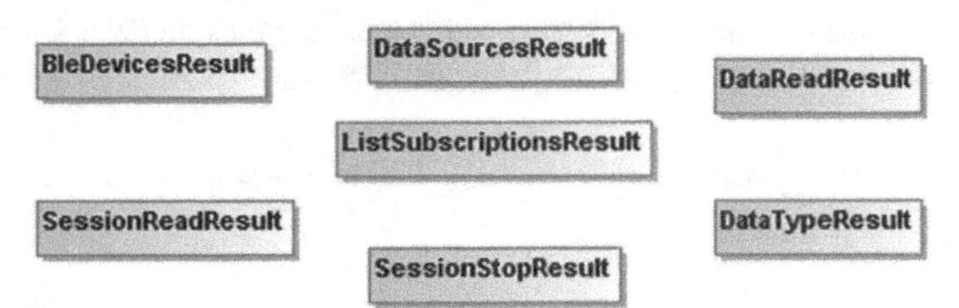

Figure 9-13A result package.

as well as the *Parcelable* interface, though this detail has not been depicted in the diagram.

9.5.1 BleDevicesResult

The *BleDevicesResult* class represents the result of a call to the *BleApi*'s *listClaimedBleDevice* method. The *BleDevicesResult* class, like the other classes in the *result* sub-package, implements both the *Result* interface from the *Google Play Services* API's *common.api* sub-package and *Parcelable* from the *android.os* package.

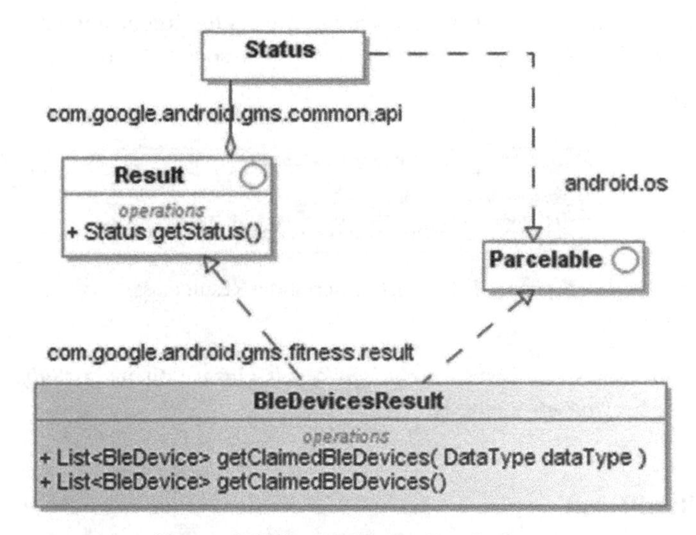

Figure 9-13B BleDevicesResult class.

Figure 9-13B shows the *BleDevicesResult* class and its overloaded *getClaimedDevices* methods. It also has methods such as *getStatus* from the *Result* interface that it implements. *BleDevicesResult* implements both the *Result* and *Parcelable* interfaces, as do all the classes in this package. However, this detail has not been repeated in subsequent diagrams of result classes.

9.5.2 DataSourcesResult

The *DataSourcesResult* class is associated with the *SensorApi*'s *findDataSources* method and the *DataSourcesRequest* class from the *request* package.

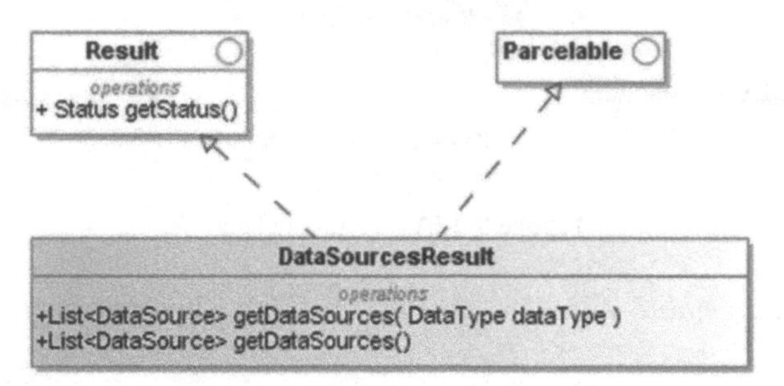

Figure 9-13C DataSourcesResult class.

Figure 9-13C shows the *DataSourcesResult* class and its overloaded *getDataSources* methods.

9.5.3 ListSubscriptionsResult

The *ListSubscriptionsResult* class has a correlation with the RecordingApi's *listSubscription* method as well as *Subscription*, which in turn is associated with the data source and data type.

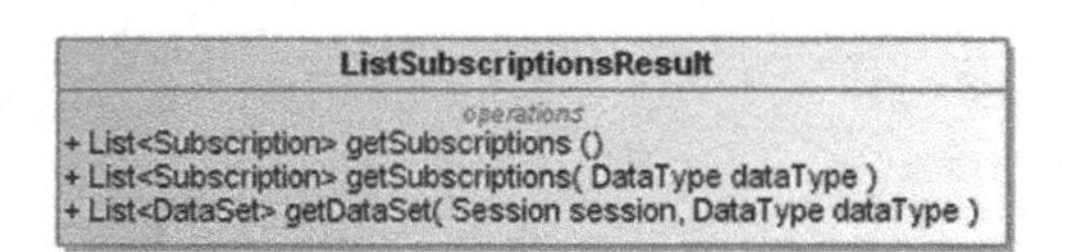

Figure 9-13D ListSubscriptionsResult class.

Figure 9-13D shows the *ListSubscriptionsResult* class with its *getSubscriptions* and overloaded *getSubscriptions* methods.

9.5.4 DataReadResult

The *DataReadResult* is associated with the *HistoryApi*'s *readData* method as well as the *DataReadRequest*.

Figure 9-13E shows the methods available in the *DataReadResult* including the overloaded *getDataSet* method. The call to *getBuckets* will be relevant if the original request had specified a bucketing strategy via any of the *bucketBy* methods available in the *DataReadRequest.Builder*. The *getBuckets* method will return an empty list in case there is no data available or if a failure is encountered.

Figure 9-13E DataReadResult class.

9.5.5 SessionReadResult

The *SessionReadResult* is associated with the *SessionsApi*'s *readSession* method and *SessionReadRequest*. The *SessionReadResult* contains the sessions and associated data that matched the criteria that were specified while building the *SessionReadRequest*.

Figure 9-13F SessionReadResult class.

Figure 9-13F shows the *SessionReadResult* class including its *getSessions* method and the overloaded *getDataSet* methods. The *DataSet* in turn contains *DataPoint*s, which we covered in the earlier sections.

9.5.6 SessionStopResult

The *SessionStopResult* is associated with the *stopSession* method available in the *SessionsApi*. The *stopSession* method requires the session identifier as a *String* parameter. There is no request class that correlates with the *SessionStopResult*.

Figure 9-13G SessionStopResult class.

Figure 9-13G shows the *SessionStopResult* class and its *getSessions* method that returns a list of sessions that were stopped.

9.5.7 DataTypeResult

The DataTypeResult class is associated with the *ConfigApi*'s *readDataType* method, which is useful for retrieval of shareable data types or private, custom data types defined in your App. The *readDataType* method accepts a String parameter with the name of the data type. There is no corresponding request class associated with the *DataTypeResult*.

Applications require user permission in order to access a shareable data type. In case the application is missing such permission, it will need to address the received status code of *FitnessStatusCodes.NEED_OUTH_PERMISSIONS* by using the *Activity*'s *startResolutionForResult*, which will get the Android platform to start the appropriate *Intent* that will solicit the user's consent.

Just in case an App attempts to read some other App's custom data type—which happens to be private to the other App—the calling App will encounter an error status code of *FitnessStatusCodes.INCONSISTENT_DATA_TYPE*. There is no resolution for this because it is not meant to be allowed that the custom data types defined by one App are accessed by another App.

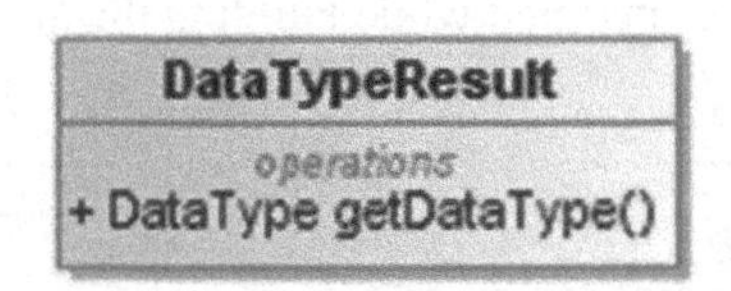

Figure 9-13H DataTypeResult class.

Figure 9-13H shows the *DataTypeResult* class that has a *getDataType* method. The *getStatus* method that the *DataTypeResult* class implements by virtue of implementing the *Result* interface has been shown only in the initial diagrams in this result series and particularly in Figure 9-13B (*BleDevicesResult*) in some detail.

9.6 service sub-package

The *Google Fit* platform has built-in support for local hardware fitness sensors on Android handheld devices and *Android Wear* devices. The *Google Fit* platform also has support for peripheral *Bluetooth* **LE** *(Smart)* devices that support standard GATT profiles. *Bluetooth LE* is an interconnectivity technology that is widely used and particularly pertinent to fitness sensors and peripheral devices. *Bluetooth LE* however is not backward compatible with the "standard" or "classic" *Bluetooth* technology. And both classic *Bluetooth* and *Bluetooth LE* are themselves only one of several interconnectivity technologies for peripherals that are available today. Thus, the *Google Fit* platform currently provides off-the-shelf support for fitness peripheral devices, limited to those that support *Bluetooth LE*-based interconnectivity.

You may just possibly have the need for your *Google Fit* App to work with non-*Bluetooth LE* fitness sensors (the most obvious example of this is classic *Bluetooth* fitness sensors). Although this happens to be a scenario that the *Google Fit* platform does not address off the shelf, the *service* package provides the APIs that can help your App work with a sensor that does not support *Bluetooth LE* and expose it to the *Google Fit* platform. Once exposed as a software sensor-based *Google Fit* sensor, from that point forward, it can be accessed via the standard *SensorsApi*.

In another scenario, your application's algorithm can analyze and interpret raw sensor data from an Android device's accelerometer or image data from an Android device's camera, in order to implement a software-based step counter or a heart rate monitor, respectively. With the human finger placed on a phone's camera for instance, the image data captured over a period of say 1 minute can be analyzed and interpreted to determine a heart rate reading—microscopic movements of the finger cause patterns of color changes in a repeating cycle, the periodicity of which if interpreted correctly will tend to coincide with the heart rate. We have covered computer vision-based sensors in Chapter 2. The raw data in such cases does not directly constitute fitness data, but the software-based algorithm can analyze and interpret such raw data to compute or infer fitness data readings such as the heart rate and more. This is another scenario that the *Google Fit* platform does not support off the shelf; however, *service* package's API can be useful to expose such software application-based sensor data to the *Google Fit* platform.

Thus, the *service* sub-package will be useful for implementing third-party software-based sensors that are exposed to and compatible with the *Google Fit* platform in scenarios such as:

1. You would like your *Google Fit* App to support a fitness sensor hardware device that uses a connectivity technology other than *Bluetooth LE* and expose such data to the *Google Fit* platform.
2. You would like to write a software-based sensor that uses more fundamental, raw sensor, or image data to compute fitness data and expose such data to the *Google Fit* platform.

Chapter 5 covered several interconnectivity and discovery technologies that have a correlation with implementing custom software sensors. Just in case you had skipped that chapter and are interested in exploring implementing a custom third-party *Google Fit* sensor, now is a good time to revisit Chapter 5.

If you already have live sensor data that is compatible off the shelf with the *Google Fit* platform, you will be less likely to need to implement your own software-based sensor.

Figure 9-14 shows the *FitnessSensorService* and *FitnessSensorServiceRequest* classes and the one interface in the service sub-package.

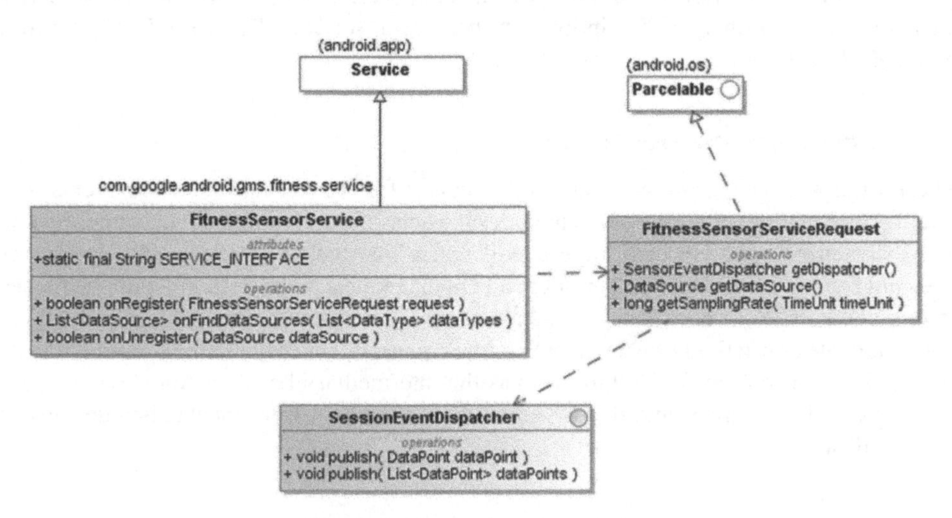

Figure 9-14 service sub-package.

9.6.1 FitnessSensorService

The *FitnessSensorService* helps your application to expose a software-based sensor to the *Google Fit* platform. After that, other Apps can claim the software-based sensor "device" and use it via the *SensorsApi*'s standard interfaces.

The *FitnessSensorService* is abstract and extends the abstract *android.app.Service*. Your application's service implementation will need to extend *FitnessSensorService* and implement the abstract methods including *onFindDataSources*, *onRegister*, and *onUnregister* from the *FitnessSensorService* hierarchy as well as the *onBind* and life cycle methods from the *Service* hierarchy. Registration is on the basis of the data source. As long as there is an active registration for a data source, your service must publish *DataPoints* at the requested sampling rate and batch interval using the dispatcher.

Your service will naturally need an entry in the application's *AndroidManifest.xml*. The *exported* flag will need to be set to ***true*** as it is expected to be exposed to the *Google Fit* platform and interact with external Apps. Furthermore, the service's entry in the manifest will need to declare a mime type filter based on the *Google Fit* data type (standard or shareable) that it supports. Below is a snippet of such an entry in the *AndroidManifest.xml* file:

```
<service android:name="io.wearbook.fitness.HeartrateSensorService"
        android:exported="true">
  <intent-filter>
    <action android:name="com.google.android.gms.fitness.service.FitnessSensorService" />
    <data android:mimeType="vnd.google.fitness.data_type/com.google.heart_rate.bpm" />
  </intent-filter>
</service>
```

The *Google Fit* platform will bind to your service and remain bound as long as there is an active registration or subscription for the sensor service. The *Google Fit* platform manages the life cycle of your service.

9.6.2 FitnessSensorServiceRequest

The *FitnessSensorRequest* encapsulates the request for registering for sensor events from your service implementation, which includes the data source, batch interval, sampling rate, and dispatcher to publish data to. You will notice that the *FitnessSensorServiceRequest* does not have any Builder and does not reside in the request package along with the request family classes that we covered earlier. Apps that are interested in receiving sensor data do not instantiate or use the *FitnessSensorServiceRequest*, rather they use the *SensorApi* in the standard way. The *Google Platform* acts as the intermediary between Apps that desire to subscribe to data sources, and this is achieved via the intent-filter mechanism declared in the manifest.

9.6.3 SensorEventDispatcher interface

The *SensorEventDispatcher* is used by the *FitnessSensorService* implementation to dispatch or push events out to the *Google Fit* platform. Your implementation of the *FitnessSensorService* is the intermediary between the sensor and the *Google Fit* platform. The *publish* methods in the *SensorEventDispatcher* help your service publish individual *DataPoint*s as well as a batch of *DataPoint*s.

References and Further Reading

https://developers.google.com/fit/

https://developers.google.com/fit/android/reference

http://developer.android.com/reference/com/google/android/gms/fitness/package-summary.html

https://developers.google.com/fit/android/data-type

http://tools.ietf.org/html/rfc6749

http://en.wikipedia.org/wiki/Value_(computer_science)

http://developer.android.com/reference/com/google/android/gms/common/api/Result.html

http://developer.android.com/reference/com/google/android/gms/common/api/Status.html

Part V Real-World Applications

This section has one short chapter that provides an overview of the possibilities for useful Wearable applications in the real world, from a long-term and broad perspective.

Wearable Android™: Android Wear & Google Fit App Development, First Edition. Sanjay M. Mishra.
© 2015 John Wiley & Sons, Inc. Published 2015 by John Wiley & Sons, Inc.

Chapter 10 Real-World Applications

10.1 Real-World Applications

Wearable devices and applications are not intended to increase the amount of information that a user needs to consume, rather they are meant to make the consumption of such information easier. Given that there are tasks and actions that a user needs to carry out, Wearable devices and applications can it make more convenient for the user to do so.

10.2 Handheld Application Extension

Given an existing handheld application, there will likely be a small subset of the functionality that makes a compelling case for extension into the Wearable platform. Probably, the most important and time-sensitive notifications in the handheld application represent what will likely be beneficial for extending to the Wearable flavor of the application.

10.3 Home Automation

Many home appliances and accessories such as washers, dryers, toasters, thermostats, light bulbs, and so on have commenced to include network connectivity via Bluetooth, Wi-Fi, and so on. Home appliance manufacturers have commenced to recognize the advantages of making their devices network enabled and service oriented—based on inter-operable and

Wearable Android™: Android Wear & Google Fit App Development, First Edition. Sanjay M. Mishra.
© 2015 John Wiley & Sons, Inc. Published 2015 by John Wiley & Sons, Inc.

open standard-based technologies, rather than proprietary mechanisms. Wearable and other IoT devices in the consumer arena can tend to feed off one another and provide incremental value to the consumer, by their ease of inter-interoperability.

The presence of more "smart" networked devices in the consumer's network brings up the opportunity for interaction and control via Wearable/smart watch-based applications. Consumers tend to appreciate being able to control their home appliances via their Wearable/smart watches and receive relevant notifications as well.

10.3.1 Home Entertainment

Several entertainment, media, video, audio systems, and accessories have various forms of connectivity such as wired or wireless connectivity, Bluetooth, and so on. Wearable applications that interact with and facilitate control over entertainment systems can be very useful.

10.3.2 Gaming

Wearable devices such as smart watches with sensor-based applications have potential in gaming, as do accessories like Wearable bands and vests with sensors. Smart watches with relevant sensors create opportunity for Wearable applications that transform a generic smart watch to a game console by detecting and processing motion, orientation, balance, and so on. Consumers tend to prefer a generic hardware device, which in conjunction with quality software applications transforms the generic device into a specialized gaming device for the duration that they are engaged in playing games. This also tends to make the game more available and lowers the overall price of the game.

10.4 Wearables at the Workplace

Wearables at the work have many obvious use cases. For one, busy executives need to remain focused on real-world activities, such as a business meeting, while also keeping up to date about other important updates (with minimal overhead, via glanceable interactions). Glancing at a watch is less intrusive and more polite.

In case of mobile field staff who work with tools and/or handle real-world workloads, there are many scenarios where the Wearable can prove more useful than a handheld device—via glanceable information updates and action-based interactions for acknowledgments and statuses.

10.5 Fitness, Health, and Medical

The medical "triage" of heart rate, temperature, and blood pressure, which are typically measured at a doctor's office or hospital, can be measured by consumers themselves as part of their fitness data—at various times of the day and upon engaging in different activities. Although *Google Fit*, for instance, currently excludes medical applications, in the long term, there are likely to be changes to the laws that are more accommodating toward consumer-based devices and applications that serve a function that overlaps with formal medical sensor devices and applications. While the cost of health care has been rising, the cost of consumer electronics and generic software that can measure certain health-related parameters is falling. Many manufacturers of fitness sensor devices also happen to manufacture medical devices. With more consumers interested in, and with the ability to measure and store their own fitness parameters via cost-effective means, it is possible and even likely that these two worlds (of formal medical records and consumer's own fitness records) will not remain isolated for very long. This opens the door to innovation and opportunity in the arena of fitness, health, and medical applications.

10.5.1 Predictive and Proactive Consumer Health

In the long run, the collection of fitness and health data has the potential to help provide a proactive and predictive approach toward health care and management. With many parameters such as weight, heart rate, blood pressure, etc. collected routinely on a periodic and frequent basis, there is opportunity to analyze their trends over a period of time and gain some useful insights.

Advances in nano-technology are beginning to make it practicable for ingested probe pills/nano-bots to probe the human body for diagnostic information and communicate with their Wearable and handheld devices to collect advanced diagnostic data, while users go about their normal daily schedules. Using both routine and advanced sensor data, in conjunction with data analysis and prediction algorithms, it is likely that doctor's visits will be based upon dynamic recommendations rather than on some fixed, periodic schedule.

10.5.2 Wearables for Medical Professionals

Head-mounted displays have been used in medicine and surgery for reality augmentation and training. These have been based on custom hardware and software, which is typically more expensive. With the arrival of consumer Wearables including *Google Glass*, there is opportunity for new applications that can provide cost-effective augmented reality solutions that address some of the use cases in the medical arena. Vision-based systems (as covered in Section 2.9.1) have introduced innovative sensing and measurement mechanisms for bodily parameters.

10.5.3 Wearables and Remote Medical Diagnostics

In some developing nations with limited resources and few medical doctors, remote monitoring using Wearable, IoT, and handheld devices is already in use today—mobile field staff deliver basic health care with remote assistance from centrally located systems, doctors, and hospitals. Sensor, handheld, and Wearable devices help in the acquisition of body parameters. Such innovation stems from absolute necessity.

In developed nations and depending on the particular country specific laws governing medical practice and standard practices, the particulars may vary. Oftentimes in developed countries, remote monitoring has been in use for patients with special needs and those needing long-term monitoring. With regard to the delivery of basic health care, many developed nations have a more conventional and conservative approach. However, health care costs keep rising, and many routine doctor's visits turn out to be unnecessary or avoidable. The leveraging of consumer wearables and generic monitoring devices for purposes of medical purposes can help reduce health care costs. It can also be aligned with the predictive and proactive approach covered in Section 10.5.1. There has been much work done in the field of mobile device-based remote diagnostics and screening, more details of which can be found at

http://miter.mit.edu/articlesana-providing-hope-healthcare-through-mobile-technology/
http://www.ncbi.nlm.nih.gov/pmc/articles/PMC3149792/

10.6 Industrial Manufacturing

Wearables have potential use in industrial manufacturing of specialized components in automobile, aerospace, and other industries. Much like in the case of devices for medical professionals, consumer Wearable devices with innovative software have the potential to provide cost-effective augmented reality solutions for industrial manufacturing.

10.7 Civic, Government, and Democracy

Wearables have been in use in law enforcement, and consumer Wearable devices with innovative software solutions have the potential to provide cost-effective augmented reality solutions in this arena. IoT devices have applicability in the management of civic infrastructure and resources.

In the representative form of government—commonly in existence today—citizens get to choose their representatives at election time; after the election is over, citizens depend on their representative to make decisions on their behalf. Direct democracy is a form of democracy in which citizens get to vote directly on all issues and policies. The growth of ubiquitous computing and technology makes it easier to implement "Direct democracy" in practice. Direct democracy allows the entire citizenry to participate in government directly to the degree that they would individually care to. Direct democracy can coexist with the representative form of government; however, the votes on issues that come straight from the citizenry directly might become difficult to ignore.

References and Further Reading

http://en.wikipedia.org/wiki/Google_Contact_Lens
http://en.wikipedia.org/wiki/Head-mounted_display
http://www.cnn.com/2015/01/29/tech/mci-nanobots-eth
http://web.mit.edu/zacka/www/moca.html
http://miter.mit.edu/articlesana-providing-hope-healthcare-through-mobile-technology/
http://www.ncbi.nlm.nih.gov/pmc/articles/PMC3149792/
http://en.wikipedia.org/wiki/Direct_democracy
http://en.wikipedia.org/wiki/Demoex
http://www.cnet.com/news/wearable-book-lets-readers-feel-the-fiction/

Index

Note: Page numbers in *italics* refer to Figures; those in **bold** to Tables.

Wearable Android™: Android Wear & Google Fit App Development, First Edition. Sanjay M. Mishra.

© 2015 John Wiley & Sons, Inc. Published 2015 by John Wiley & Sons, Inc.

Printed in the USA
K00783SCI082815 01S29053000000000695